ENGINEER
Through the year

Grades 3–5

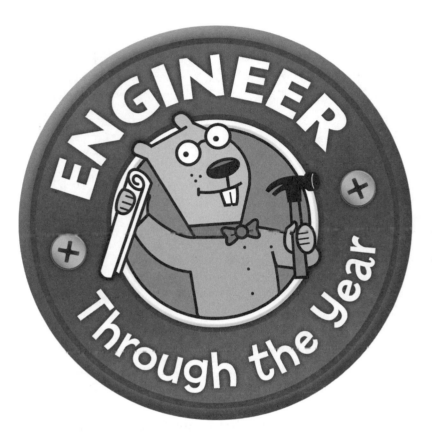

# 20 Turnkey STEM Projects
## to Intrigue, Inspire & Challenge

**Sandi Reyes**

**Crystal Springs**
SDE **BOOKS**

A division of Staff Development for Educators

**Peterborough, New Hampshire**

Published by Crystal Springs Books
A division of Staff Development for Educators (SDE)
10 Sharon Road, PO Box 500
Peterborough, NH 03458
1-800-321-0401
www.SDE.com/crystalsprings

Published 2012
Printed in the United States of America
16 15 14 13 12   1 2 3 4 5

ISBN: 978-1-935502-49-4
e-book ISBN: 978-1-935502-50-0

Library of Congress Cataloging-in-Publication Data

Reyes, Sandi, 1969-
 Engineer through the year, grades 3-5 : 20 turnkey STEM projects to intrigue, inspire &
challenge / Sandi Reyes.
     pages. cm
 Includes bibliographical references.
 ISBN 978-1-935502-49-4 (alk. paper)
 1.  Engineering--Study and teaching (Elementary)--United States. 2.  Engineering--Study and
teaching (Elementary)--Activity programs.  I. Title.

LB1594.R46 2012
620.0071--dc23

2012038759

For Gary...

No matter how outlandish my ideas or how many hours I work,
you always support me and believe in me. For this, I am blessed.
You are truly my gift from the universe.

# Contents

# Acknowledgments

No matter what project I dive into, my husband, Gary, is always there to support me. So, when I told him I was planning to go directly into writing a second book after finishing the first, his reaction, of course, was "Great!" I'm not sure his enthusiasm actually lasted throughout the months of writing and editing, but nonetheless, he never complained, regardless of how many hours I sat in front of a computer screen. So, Gary, in case I didn't say it enough during the writing, "Thank you for putting up with me and for always being willing to play along when I was stuck and needed to think out loud. Thank you for understanding that this book is something I just needed to do. And thank you for your unending love, which is what props me up when I'm ready to fall down."

The pride that my parents and my aunt have in me is something that I treasure and keep in mind with each new endeavor. Mom, Dad, and Aunt Barbara, thank you for always making me feel as if I can do anything and everything (especially when it's so obvious that I can't). When you tell me I can do it, it somehow makes it true for me and pushes me through the tough times. Your belief in me motivates me, and I drew on that when life seemed overwhelming and coming up with new ideas seemed impossible. Thank you and I love you.

To Jason, Jacquie, Tonia, Kellie, Ariana, and Maya...each of you so unique, with so much to offer the world. I'm lucky that you've allowed me into your lives. Thank you for letting me be a part of your worlds and letting me learn from you. Watching each of you grow has helped me to be a better person and teacher. Your contributions to this book are immeasurable, as I thought of each of you and your strengths every time I created a new activity.

Very few people can say they've had the same best friend for their entire lives. I am blessed to be one of them. Christy Miller knows everything about me and loves me anyway. Thank you for understanding that when I say I'm coming over for tea, what I really mean is, "I'm coming over to pick your brain and run some new engineering challenge ideas by you." For someone who knew nothing about engineering, you sure did a great job listening and offering suggestions. I am lucky to have you.

This book would not exist were it not for the incredible team at SDE and Crystal Springs Books. I can still remember the day that Lisa Bingen approached me about the possibility of writing a book. It was a "dream come true." Thank you, Lisa, not only for recognizing the importance of this topic, but also for seeing something in me and having such faith in my ability. Lisa couldn't have made me that offer if Melissa Sheldon hadn't given me the opportunity to speak at SDE conferences. Thank you, Melissa, for believing in me and giving me a chance. I am forever indebted to you. I just don't know how to begin to thank Sharon Smith for all the wisdom and guidance shown me throughout the process of writing not only one book, but now two. Somehow, things that seemed murky in my mind became so much clearer after we would talk. Sharon, you have an amazing talent for helping me see through the clouds to the heart of the matter. And for that, I'll be forever grateful.

Getting inside my head and actually figuring out what is going on there is no easy task, yet Marianne Knowles seems to be able to do it effortlessly. I could write a page of "thank you" and it wouldn't be enough to express my appreciation of how you've been such an amazing partner on this journey. You are a brilliant editor and a superb person. I am better off for having known you, and this is a better book for having had you as its editor.

# Introduction

When I was asked to incorporate engineering education into my classroom, I wondered how I could possibly squeeze any more into the school day, not to mention my workload. Then I compared what I was doing with my students to the requirements of engineering education and to STEM education in general. I was pleasantly surprised to find that much of what was already happening in my classroom—and probably in your classroom too—is exactly what STEM education is all about.

The best part was that I could take some existing curricular concepts and add an engineering challenge to each of them, rather than rewriting the curriculum or adding yet another topic to the school day. By combining the science and math I was already teaching with a real-world design challenge, I was actually making less work for myself and better differentiated instruction for my students—all while giving them the benefits of STEM education and 21st century skills development.

## What Is STEM? Why Is It Important?

STEM stands for Science, Technology, Engineering, and Mathematics. The essence of STEM is problem-based learning. Children are presented with a problem and design something that solves the problem. In the process, they activate science knowledge and math skills. They use technology to research, design, test, and present their ideas. As they work together, students learn and practice good collaboration and verbal communication—key skills identified by The Partnership for 21st Century Skills.

> "STEM jobs are the jobs of the future. They are essential for developing our technological innovation and global competitiveness."
>
> —from "STEM: Good Jobs Now and for the Future," U.S. Department of Commerce, Economics and Statistics Administration

## What Does the Research Say?

Several studies have been conducted over the past 20 years regarding both STEM and project-based learning. Here are two of the most impressive.

**"Open and Closed Mathematics":**
Dr. Jo Boaler conducted an investigation into two schools in England that had similar students, but different approaches. After three years, students who followed a traditional approach developed a procedural knowledge that they had difficulty applying in unfamiliar situations. Students who learned mathematics in an open, project-based environment developed a conceptual understanding that allowed them significantly higher achievement in a range of assessments and situations.
                              —Boaler, "Open and Closed Mathematics"

***How People Learn***:
A few key points out of a wealth of research on how to best teach students: Students are more engaged and achieve better overall results when working in small collaborative groups on tasks that allow application of skills and processes. The more choice students have in their learning, the more they learn. And, the teacher who gradually releases responsibility to her students will see great results.
                —Bransford, John D., et. al., *How People Learn*

1. Investigate
2. Brainstorm
3. Plan
4. Build
5. Test & Present

## The Engineering Design Process

If you do any research, you'll find several different versions of the engineering design process. Most have been created to describe the general process followed by adults and then scaled back for use with children. The engineering design process used in this book, however, was designed for use in elementary schools. It intentionally follows a path similar to the one used to describe the scientific inquiry method. This was done to make the design process fit naturally into existing classroom practice; it is also developmentally appropriate for students understand and use.

That being said, it is important to recognize that engineering design, like scientific inquiry, is messy. The engineering design diagram may show a perfect sequence, with one stage of the process leading neatly into the next, but this rarely happens in real life. Engineers, whether children or adults, bounce around, hitting each stage in an order that makes sense for a particular person and a particular design.

Certain behaviors are hallmarks of each stage of the process, regardless of the order in which the stages occur. The next several pages describe each stage as it's used in this book.

## Young Learners Are Natural Engineers

You may be thinking, "Engineering with children? They don't know a thing about how things work in the real world!" It may sound daunting, but it's really not difficult to do. From the time their hands can first grasp objects, children build. They put things together and take things apart. They may ask how things work, but they do not always accept explanations—they investigate to find out for themselves. When faced with a challenge, children look to solve it without the restrictions that we adults place on ourselves. Nothing is impossible in a child's mind, and it is this kind of thinking that allows children to create ingenious inventions. It is what makes children such natural engineers.

This book offers step-by-step engineering activities that correlate to the most common yearlong themes taught in the elementary grades. Twenty Challenges are provided—enough for two options for each month of the school year. Each activity lists suggested curricular connections for seamless implementation into your current curriculum; these are summarized in a chart on page 11. The Content & Skills Alignment Charts on pages 183–187 show the integration of all 20 engineering Challenges with the Next Generation Science Framework, 21st century skills as identified by The Partnership for 21st Century Skills, and the Common Core Standards for Mathematical Practice.

# How to Use This Book

Each engineering design Challenge is organized in the same easy-to-follow format. But don't let the structure mislead you. The Challenges themselves are not a set of recipes or list of how-to instructions. They are open-ended experiences that require students to stretch themselves as they learn to generate ideas, solve problems, and work collaboratively with peers.

## Challenge

Twenty Challenges are available— enough to let you choose the ones that best mesh with your curriculum and schedule throughout the year.

## Curriculum Connections

The Curriculum Connections offer ideas of where to link a Challenge to topics that are commonly taught at the upper elementary grades.

## Criteria for Product; Constraints for Challenge

The Criteria list clearly states for you and your students how to determine that a solution is successful. The Constraints specify limitations on time, materials, budget, or other resources.

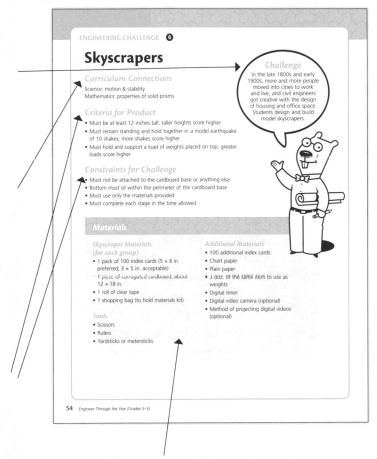

## Materials

Quickly scan the Materials lists to determine which materials and tools you already have and which you may need to collect. Quantities are listed either for a group or for the whole class, as noted. Any special or optional equipment for the Challenge is listed under Additional Materials.

## Before You Begin

This handy list provides directions for materials preparation and advance planning.

### 1 Investigate

Sessions 1 and 2
Investigate contains two teacher-led activity sessions; these provide students with guided experiences for materials exploration and research. During Investigate, students review relevant science and math concepts, explore examples of the objects they will design, examine the available materials, and review the Criteria for Product and Constraints for Challenge.

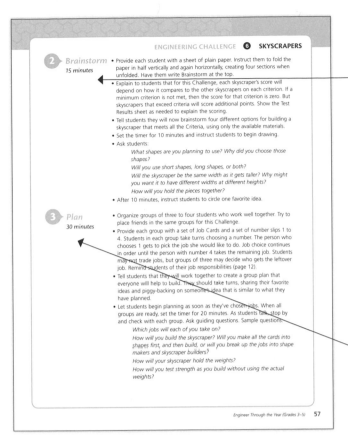

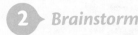

**2** *Brainstorm*
15 minutes

- Provide each student with a sheet of plain paper. Instruct them to fold the paper in half vertically and again horizontally, creating four sections when unfolded. Have them write Brainstorm at the top.
- Explain to students that for this Challenge, each skyscraper's score will depend on how it compares to the other skyscrapers on each criterion. If a minimum criterion is not met, then the score for that criterion is zero. But skyscrapers that exceed criteria will score additional points. Show the Test Results sheet as needed to explain the scoring.
- Tell students they will now brainstorm four different options for building a skyscraper that meets all the Criteria, using only the available materials.
- Set the timer for 10 minutes and instruct students to begin drawing.
- Ask students:
  *What shapes are you planning to use? Why did you choose those shapes?*
  *Will you use short shapes, long shapes, or both?*
  *Will the skyscraper be the same width as it gets taller? Why might you want it to have different widths at different heights?*
  *How will you hold the pieces together?*
- After 10 minutes, instruct students to circle one favorite idea.

**3** *Plan*
30 minutes

- Organize groups of three to four students who work well together. Try to place friends in the same groups for this Challenge.
- Provide each group with a set of Job Cards and a set of number slips 1 to 4. Students in each group take turns choosing a number. The person who chooses 1 gets to pick the job she would like to do. Job choice continues in order until the person with number 4 takes the remaining job. Students may not trade jobs, but groups of three may decide who gets the leftover job. Remind students of their job responsibilities (page 12).
- Tell students that they will work together to create a group plan that everyone will help to build. They should take turns, sharing their favorite ideas and piggy-backing on someone's idea that is similar to what they have planned.
- Let students begin planning as soon as they've chosen jobs. When all groups are ready, set the timer for 20 minutes. As students talk, stop by and check with each group. Ask guiding questions. Sample questions:
  *Which jobs will each of you take on?*
  *How will you build the skyscraper? Will you make all the cards into shapes first, and then build, or will you break up the jobs into shape makers and skyscraper builders?*
  *How will your skyscraper hold the weights?*
  *How will you test strength as you build without using the actual weights?*

## Helpful Classroom Technology

The use of current technology in the classroom both enhances teacher instruction and develops student skills. Note, however, that students will be able to complete all design Challenges even if certain tools are not available. The following are recommended for use in some Challenges.

### Teacher Use (Recommended)

- Document camera
- Computer
- Projector or display monitor
- Internet access
- Digital camera (still)
- Digital camera (video)

### Student Use (Optional)

- Digital cameras (still)
- Digital cameras (video)
- Word processing software
- Presentation software (for example, PowerPoint)
- Computers
- Internet access

**2** *Brainstorm*

During Brainstorm, students imagine what some solutions to the Challenge might look like and then sketch, rearrange, and get ideas onto paper without stopping to analyze the merits of each. Grouping for Brainstorm varies; this stage may be completed alone, in small groups, or as a whole class, depending on the Challenge.

**3** *Plan*

Students work in groups of three to four to develop the collaboration skills that rank high in importance in STEM and in 21st century skills education initiatives. Each student in a group chooses, or is assigned, one of four group jobs: Speaker, Timekeeper, Materials Manager, or Recorder; job assignments change with each Challenge. During Plan, students share and evaluate the ideas generated during Brainstorm and then create one agreed-upon design.

The discussions that take place during Plan will reveal to you what your students know and how they think. It is during this time that connections to prior learning experiences take place, and that students are challenged to interpret the causes and effects of design decisions.

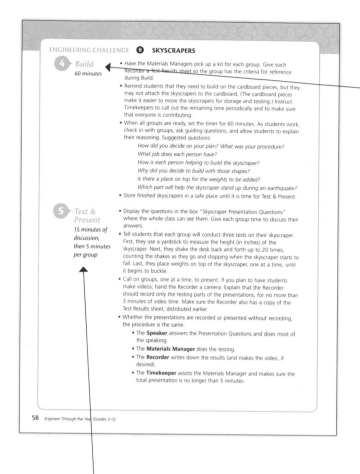

## 4 Build

Build is the most active stage of the engineering design process. Students put materials together and work cooperatively to build the objects they have designed. All students take part in building and decision making, as well as performing the group jobs assigned during Plan. Students frequently test their solution during Build to determine whether it meets the Criteria for Product.

Students start Build by following their plan exactly, but challenges inevitably arise, especially during testing. Making changes to the original plan is a necessary part of the engineering design process. Just as important is the need for all group members to agree to the changes, and for the group to get approval for, and record, the changes being made.

## 5 Test & Present

In any real-world design task, engineers and scientists must present and demonstrate their work. The audience may include colleagues, investors, executives, politicians, the public, or consumers. Every design Challenge offers an opportunity for students to present their solutions and test results to an audience. To mimic real-world presentations, a variety of audiences and presentation styles are suggested.

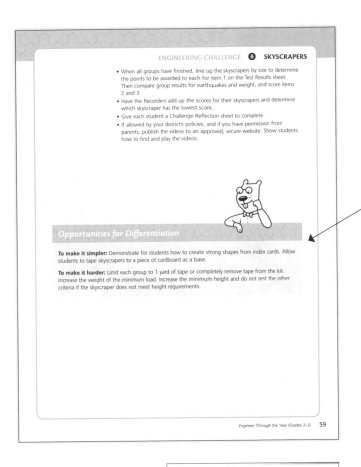

## Opportunities for Differentiation

Options for making each design Challenge simpler or harder let you adjust the classroom experience according to your students' grade level and prior experience with project-based learning, or to suit the time you have available to complete a project.

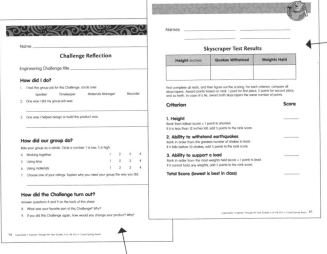

## Test Results

Challenge-specific Test Results sheets give student groups a place to record results and evaluate their product. Test Results, and any additional Challenge-specific copymasters, are provided at the end of each engineering design Challenge.

## Challenge Reflection

The Challenge Reflection sheet helps individual students to reflect on their experience in engineering and in functioning as part of a collaborative group. Students use the same set of questions to evaluate each Challenge, which provides the opportunity to compare experiences throughout the school year.

# Moving Through the Year

Each Challenge is set up in exactly the same format to create an easy-to-follow structure. Within this set structure, however, the directions for implementing each stage of the design process vary. The different strategies offered for each stage of a Challenge give students the opportunity to practice a variety of skills and to apply skills in new ways. The Challenges offer ways to:

- guide student investigation
- support brainstorming
- group students
- choose student jobs for group work
- offer materials for building
- present results
- record observations and reflections

Strategies progress through Challenges 1 to 20 from basic to complex and from teacher-directed to student-directed. The Challenges release more responsibility to students as they become more capable of managing the design process on their own.

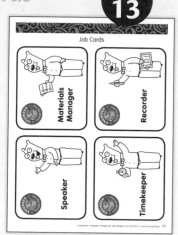

**Using Job Cards**

**Page 13**

Before beginning the Challenges, make a set of four Job Cards for each group, working from the copymaster on page 13. Students use these cards to select jobs randomly, or to keep track of which job they have when jobs are assigned another way.

You're probably wondering if this means that you must start at the beginning of the year, do every Challenge, and follow the guide exactly as it is written. Fortunately, the answer is "No." There are many ways to work with this guide and adjust it to the needs of your classroom.

## Starting Midyear

If you begin to use this book in the middle of the year, or if you choose to do only a few activities throughout the year, then your students may not be ready for the suggested strategies for grouping or developing materials lists in a particular Challenge. In this case, browse back to Challenges in the lower numbers to find strategies that will work for your class. Think of it as a mix-and-match lesson plan design process. You may choose an activity based on its Curriculum Connection being relevant to your current studies, and then borrow strategies from other Challenges for certain stages of the design process.

# Planning Your Time

Another aspect of planning is figuring out how to fit all the stages of the engineering design process into your current daily schedule. There are natural places to stop each day as you work through a design Challenge. Not all activities have the same time estimates for each stage of the process, but the chart below offers a basic look at how the process divides up.

| | |
|---|---|
| **Before the Challenge** | Time needed to complete the preparation listed in **Before You Begin** varies depending on the challenge. |
| **Day 1** | The **Investigate** stage has two sessions, each of which can be completed in a day or less. |
| **Day 2** | The **Brainstorm** and **Plan** stages can usually be completed on the same day, with careful attention to time management and minimal issues with transition. |
| **Days 3 & 4** | **Build** will take at least one session, but plan on two, depending on time estimates provided for the design Challenge. |
| **Day 5** | **Test & Present** typically takes 5–10 minutes per group and can be completed in one session. More involved presentations that require special testing setups or creating digital media may take two sessions. |

*This chart assumes a one-hour session per day. Plan for additional days if your sessions are less than one hour.*

Regardless of whether it is the beginning of the year, the end of the year, or anywhere in between, you can introduce your students to engineering. Choose a Challenge that supports where you are in your curriculum and get started!

# Adapting Challenges for Multiple Grades

Because this book is meant to work in a grade-level span, it is possible that teachers of different grade levels within a building will want to use the same Challenges, resulting in fifth-grade students being presented with the same Challenge they completed in third or fourth grade. If this happens, review the "make it harder" options under Opportunities for Differentiation. Other ways to make familiar Challenges new include:

- If there is no budget limit included in the Challenge, add one.

- Add a new length, width, height, volume, or weight requirement to the Criteria for Product.

- Think of something more that the object needs to do and add it to the Criteria for Product.

- Require the product to be visually attractive, if this is not part of the challenge.

- Require students to perform "market research" during Investigate. Have them gather information from potential "customers" to find out what they expect from the product.

- Make Test & Present more challenging by adding specific technology requirements, such as the creation of brochures, newspaper ads, PowerPoint presentations, videos, or spreadsheets.

## A Word About Safety

The engineering design Challenges in this book are safe for students in grades 3 through 5, but any activity carries some risk, however small. In general, the procedures you follow for safety during your regular science lessons also apply to engineering Challenges. You can help further reduce the chance of mishaps by following a few commonsense guidelines.

- Be aware of food allergies among your students and substitute materials as needed.

- Check your state's guidelines before bringing materials from nature into the classroom.

- Any tool that has a sharp edge or that generates heat should be handled only by the teacher or an adult aide. (Examples: glue guns, pen knives, and awls)

- Use electrical appliances at least 3 feet away from any source of water. Avoid running electrical cords across pathways where students need to walk.

For further information, refer to the resource links on The National Science Teachers Association safety portal (www.nsta.org/portals/safety).

# Curriculum Connections

Every Challenge connects to topics that may already be in your regular curriculum. The topics are identified as Curriculum Connections on the first page of each Challenge and are summarized in the chart below.

| Challenge | Curriculum Connections |
|---|---|
| 1 Rockets | **Science:** forces & motion, potential (stored) & kinetic energy, variables |
| 2 Catapults | **Science:** forces & motion, potential (stored) energy |
| 3 Compasses | **Science:** magnetism; **Social Studies:** explorers |
| 4 Healthy Snacks | **Mathematics:** measurement, fractions; **Health:** nutrition |
| 5 Simple Shelters | **Science:** natural resources; **Social Studies:** Native Americans; people, places, & environments |
| 6 Skyscrapers | **Science:** motion & stability; **Mathematics:** properties of solid prisms |
| 7 Model Cars | **Science:** motion, speed, variables; **Mathematics:** measuring distance & time, comparing numbers, division, rates |
| 8 Board Games | **Science:** define a problem, design a solution; **Mathematics:** classifying, two- & three-dimensional shapes |
| 9 Ski Lifts | **Science:** simple machines, forces; **Mathematics:** average (mean) |
| 10 Winter Coats | **Science:** adaptations, habitats, energy transfer; **Mathematics:** graphing |
| 11 Greenhouses | **Science:** plant needs, photosynthesis, water cycle |
| 12 Bird Feeders | **Science:** adaptations, needs of living things |
| 13 Parachutes | **Science:** properties of air, gravity; **Mathematics:** graphing |
| 14 Bridges | **Science:** net forces, Newton's third law of motion |
| 15 Land-Reuse Models | **Science:** human impact on the environment; **Mathematics:** average (mean) |
| 16 Mining Tools | **Science:** natural resources, earth materials; **Social Studies:** California & Alaska gold rushes, mining |
| 17 Oil-Spill Cleanups | **Science:** natural resources, environmental hazards |
| 18 New Businesses | **Mathematics:** operations; Social Studies: economics |
| 19 Solar Water Heaters | **Science:** energy transfer, alternative energy sources; **Social Studies:** people & the environment, Earth Day |
| 20 Roller Coasters | **Science:** potential (stored) & kinetic energy, variables |

# Group Job Descriptions

## Speaker

**Speakers** make sure everyone gets a chance to talk. They also help solve disagreements, bring group questions to the teacher, arrange trades with other groups (when the Challenge allows), and answer questions when the group presents results.

## Materials Manager

**Materials Managers** get materials for the group. They also make any trades that the Speaker sets up, keep track of the budget if there is one, make sure that everyone in the group is using materials wisely, and test the product. Lastly, they organize cleanup by assigning tasks to each group member.

## Timekeeper

**Timekeepers** keep track of the time for each stage of the engineering design process. Timekeepers may assign different people to different tasks during Build so that everything gets done on time. They let the group know how much time is left, and help everyone stay on task so the work gets done before time runs out.

## Recorder

**Recorders** write down or draw all of the group's work, or direct others to do so when recording is a group effort. Recorders also keep track of any changes made to the design plan during Build, and record the results during Test & Present. Recorders take photos and videos of the group when cameras are used.

# Job Cards

**Materials Manager**

**Recorder**

**Speaker**

**Timekeeper**

Name _____

# Challenge Reflection

Engineering Challenge title _____

## How did I do?

1.  I had this group job for this Challenge. (circle one)

    Speaker          Timekeeper          Materials Manager          Recorder

2.  One way I did my group job was:

    _____

3.  One way I helped design or build the product was:

    _____

## How did our group do?

Rate your group as a whole. Circle a number: 1 is low, 5 is high.

4.  Working together                          1      2      3      4      5

5.  Using time                                1      2      3      4      5

6.  Using materials                           1      2      3      4      5

7.  Choose one of your ratings. Explain why you rated your group the way you did.

    _____

## How did the Challenge turn out?

Answer questions 8 and 9 on the back of this sheet.

8.  What was your favorite part of this Challenge? Why?

9.  If you did this Challenge again, how would you change your product? Why?

# Rockets

## Curriculum Connection

Science: forces & motion, potential (stored) & kinetic energy, variables

## Criteria for Product

- Must have a body tube, nose, and fins
- Must be 6 to 12 inches long
- Must weigh 1 pound or less, including launch stand and propellant
- Must be propelled by something other than direct human force
- Greater flight distances score higher, for both horizontal and vertical distances
- Shorter flight times score higher

### Challenge

Rockets send satellites into orbit and launch probes beyond Earth to study space. Closer to Earth, some rockets launch fireworks! Students design, build, and test model rockets.

## Constraints for Challenge

- Must use only the materials provided
- Rocket launch must follow safety guidelines
- Must be launched at a 45-degree angle (approximate)
- Rocket must not be propelled by any flammable fuel
- Must complete each stage in the time allowed

## Materials

### Rocket Materials (for each group)

- 1 empty plastic soda bottle, 2-L size
- 3 ft. of masking tape
- a few pencils or craft sticks
- 2 balloons of the same size and shape
- 2 seltzer-type antacid tablets
- 5 rubber bands
- 5 straws
- 5 sheets of plain paper
- 2 sheets of construction paper
- 5 index cards, 3 × 5-in.
- 1 ft. of waxed paper

- 1 film canister or similar plastic container with snap-on lid
- 1 shopping bag (to hold materials kit)

### Tools

- Scissors
- Hole punches
- Spoons, any type
- Paper or plastic cups, any size
- Medicine-size measuring cups, if available

### Additional Materials

- 2 lb. of baking soda
- 32 oz. of vinegar

- Water
- Chart paper
- Markers
- Plain paper
- Pencils
- Digital timer
- Stopwatch
- Rulers or tape measures
- Safety goggles
- Postal or kitchen scale for weighing rockets
- Computer with Internet access (optional)
- Projector for computer (optional)

**Note:** *Have extra rocket materials on hand for demonstrations.*

# Before You Begin

- Locate a suitable test site outdoors or in a room with a high ceiling, such as a gym. The site should have enough space behind the "launch pad" area that students can stand back at least 25 feet, and enough clear space in front that rockets launched from a 45-degree angle can travel 75 feet or farther. Make sure the space is available during the time you would like to test the rockets.

- Assemble a materials kit for each group, following the Rocket Materials list. Wrap masking tape around a pencil or craft stick before placing in the bag. Write the words Baking Soda, Vinegar, and Water on slips of paper and add to each materials kit, so students know that these materials are also available.

- Make a set of Job Cards (page 13) and a copy of the Test Results (page 22) for each group.

- Make a copy of the Challenge Reflection sheet (page 14) for each student.

- Set up the digital timer where students can check on it.

- Display the Criteria and Constraints where students can see them.

- Preview the Dragonfly TV video about testing model rockets and then bookmark it if you plan to use it during Investigate Session 2. Go to (pbskids.org/dragonflytv) and search "model rockets."

# 5-Step Process

### Investigate
*Session 1*
*40 minutes*

- Ask students to think about an object lying on the ground. The force of gravity is pulling it down. The force of the ground is pushing it up. Does it have to lie there forever? No? How can you make that object change its motion? Pick it up? Kick it? When you do that, you're putting a force on the object that is in a different direction from, or is greater than, the force of gravity pulling down on it. The unbalanced force you put on the object makes the object's motion change.

- Next, ask students what they have that lets them put that force on the object to move it. Guide students to remember that energy is the ability to change the motion or direction of an object. (Or, energy is the ability to do work, with work defined as a change in the motion or direction of an object.)

- Review the two types of energy: Potential energy is energy that is stored; it is available to cause changes in motion, but it isn't doing it yet. Kinetic energy is the energy an object has because it is in motion.

- Tell students that for their engineering Challenge, they will take on the role of aerospace engineers as they design and build their own model rockets. They need to store energy that can be used to make an object move. But they need to learn about ways to store energy before they can choose one method to use in their Challenge.

- Guide students through teacher-directed explorations of five ways of storing energy. Choose a few students to assist in each demonstration.

- **Seltzer tablet:** Give each student a cup of water and half of an antacid tablet. Have them drop the tablets into the water.
- **Baking soda and vinegar:** Give each student a cup with a spoonful of baking soda and a medicine-size cup of vinegar. Have them pour the vinegar into the cup.
- **Rubber bands:** Have students wear safety goggles and stand shoulder-to-shoulder facing a wall, at a distance of 5 feet from the wall. Keep all other students back at least 10 feet. Allow students to shoot a rubber band at the wall.
- **Peashooter:** Have students crumple a small piece of paper, place it in the end of a straw, and blow it out.
- **Stomp bottle:** Give each student an empty plastic 2-liter bottle. Instruct them to put the bottle on the floor, lying on its side. Have them place a crumpled piece of paper in front of the bottle's opening and then stomp on the bottle suddenly.

- During each demonstration, ask students to describe what happens. Then discuss as a class how this method could be used to store energy for a model rocket launch.

**1** **Investigate**
**Session 2**
**30 minutes**

- Tell students that for this Challenge, they will first come up with ideas on their own. Later, they will get into groups to plan, build, and launch their rockets.
- Describe and discuss the main parts of a rocket, including the nose, body tube, and fins. Explain that the propellant is whatever is used to launch the rocket.
- Show the Dragonfly TV model rocket video that you previewed earlier. The video shows the effects of changing three parts of a rocket's design: the body tube, the nose, and the fins. Instruct students to take notes about which designs result in the highest flight for the rocket.
- Define for students the terms *criteria* and *constraints*. Explain that engineers always work to certain specifications, called *criteria*, and that they work within certain limitations, called *constraints*. Sometimes they come up with the criteria themselves; other times a client comes up with the criteria.
- Review with students the Criteria and Constraints for this Challenge, as well as the time limits for Brainstorm, Plan, and Build. Show students the Test Results sheet so they understand how their rockets will be scored.
- Remind students that to get something to move, they need energy. Ask them to think about the different ways energy could be stored, which they investigated in Session 1.

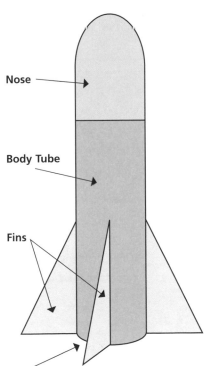

Nose

Body Tube

Fins

**Propellant applies force at base of tube**

- Remind students that they may not use their own energy to directly propel their rockets, but they may store their own energy in something else that propels the rocket.

### 2 Brainstorm
**20 minutes**

- Show students the items in a materials kit. Tell them they can use any of these materials—but no others—to design a rocket. Ask students:

    *What properties do these materials have?*

    *How can the properties of these materials help with a rocket?*

    *Which materials would be good for each part of the rocket (nose, body tube, fins, propellant)?*

- Give each student a sheet of plain paper. Have them write Brainstorm at the top.

- Ask students to work silently and independently for 15 minutes to write and draw all ideas they have for building a rocket that will travel quickly and a long distance. Explain that brainstorming is "messy thinking" and that it's more important to come up with lots of ideas than to make their pages neat. Encourage them to draw pictures and to label the parts in as much detail as possible.

- As students work, circulate around the room. If anyone is stuck, suggest starting places and ask guiding questions. Sample questions:

    *Why did you choose to use that material for that part?*

    *Which kit materials could be used for the body tube? For the fins?*

    *What are you planning to use for a propellant? How will you attach the rocket to its propellant?*

    *How will you attach the fins? The nose?*

    *What's another way you could use the same materials?*

- After 15 minutes, ask each student to choose one favorite idea and circle it.

### 3 Plan
**30 minutes**

- Explain to students that you will be dividing them into groups to plan, build, and test the rockets. Everyone in the group will help to plan and build the rocket, but everyone will also have a job to do. Explain the responsibilities of these jobs. (Similar job descriptions appear on page 12.)

    - **Speakers** help solve disagreements in the group before going to the teacher. They also bring questions and design changes to the teacher, and answer questions when the group presents.

    - **Timekeepers** keep track of the time, let everyone know how much time is left, and help keep everyone on task. (Point out the digital timer you will be setting.) Timekeepers may assign tasks as needed to be sure everything gets done on time.

    - **Materials Managers** gather materials for the group and test the product (in this Challenge, the rocket). They also keep track of the materials used and help prevent materials from being wasted.

    - **Recorders** write down or draw the group's plan, including any changes that are made to it. They also write down the score or test results.

- Break students into groups of three and four, placing students with similar rocket designs in the same groups. Include students at various academic levels in each group, whenever possible.
- Assign the jobs of Timekeeper, Recorder, Speaker, and Materials Manager by handing a Job Card to each student in a group. In groups of three, one student will need to have two jobs.
- Invite the Materials Managers to pick up one materials kit for each group. Remind students that they may look at the materials during Plan, but they should not yet start building their rockets. Explain that their kits also include three materials that are not in the bag: water, baking soda, and vinegar.
- Set the timer for 20 minutes. Tell students to go around the group, sharing their favorite ideas from their brainstorming. Ask students to discuss their ideas and decide on one group plan. Remind students to listen respectfully to one another as they share ideas.
- As students discuss, circulate around the room to offer support as needed and mediate disagreements that the Speaker can't handle. Sample questions:

    *Did everyone get a chance to show his ideas?*

    *How did you decide which ideas to use?*

    *What should you do when you don't agree?*

    *Can you combine two ideas or change an idea?*

    *What is the propellant for your rocket?*

    *What will hold the rocket at an angle for launch?*

    *Even if you don't love it, can you live with it?*

- After each group agrees on one plan, give each Recorder a sheet of chart paper. Have the Recorder draw a labeled diagram of the group's rocket design and include a detailed description.

**4** **Build**

**45 minutes**

- Set the tools, baking soda, and vinegar where students can get them.
- Tell students they will work together to build their rockets. Each group should follow its own plan, but if the group agrees they want to change their plan, the Speaker should discuss it with the teacher and get permission before the group moves ahead and builds. If plans are changed, the Recorder notes any changes on the chart paper in a different color. The Timekeeper should make sure that changes happen quickly, because no additional time will be allowed.
- As students work, circulate around the room, asking guiding questions as appropriate, but encouraging students to solve problems on their own. Sample questions:

    *How is everyone helping to build?*

    *What part of the building process are you thinking about?*

    *Is the rocket coming out the way you imagined?*

    *What do you know about stored energy that is helping you as you build your rocket?*

    *Are you able to build it exactly the way you planned?*

- Store the finished rockets in a safe place until it is time for Test & Present.

 *Test & Present*

*7 minutes per group*

- Set up the test site. Mark a 3-foot-diameter circle or similar-size square as a "launch pad." Draw an arrow on one side to show the direction for launching the rocket. On the other side, mark a line 25 feet from the launch pad as the "safety zone." On a tree, wall, or other nearby vertical surface, mark lines for High, Medium, and Low rocket heights as described on the Test Results sheet.

- Give each Recorder a Test Results sheet. Have each group measure the length of their rocket. Each group should then weigh their rocket, including everything that is needed to launch it. Recorders should note all information. (Recorders do not have to note exact weight unless you choose to have them do so; more or less than 1 pound is adequate.)

- Lead all groups outdoors with their Test Results sheets, rockets, and any items needed to launch them. Bring a stopwatch and enough safety goggles for yourself and the group that is presenting.

- Have all students stand in the safety zone. Invite each group to take a turn presenting their rocket to the class and then launching it.

- Each group comes to the launch area and puts on safety goggles. Hand the stopwatch to the Timekeeper.

- The Recorder reads the "Before the demonstration" questions.

- The Speaker presents the rocket and answers the "Before the demonstration" questions.

- All students in the group stand back from the launch pad. The Materials Manager sets up the rocket in the launch-pad area so the rocket is at an angle and pointing away from the safety zone. The Timekeeper stands in view of the launch pad. The Recorder and Speaker stand where they can compare the rocket against the marks for height.

- The whole class counts backward from five and then says, "Go!" When the Timekeeper and Materials Manager hear this, they launch the rocket and start the stopwatch. As soon as the rocket lands, the Timekeeper stops the stopwatch.

- The Recorder completes the Test Results to determine the score for the rocket. As needed, explain that the lowest possible score should be zero—scoring does not go into negative numbers. The Recorder reads the "Before the demonstration" and "After the demonstration" questions and the Speaker answers them.

- Sample questions for before and after each launch:

    **Before the demonstration:**

    *How well did your team work together?*

    *How does your rocket work? What propels it?*

    *Did you follow your original plan or make changes?*

    *What changes did you make?*

**After the demonstration:**

*Did your rocket work the way you planned?*

*What force caused the rocket to change its speed and direction of motion during launch?*

*How high did your rocket go? How far?*

*For how many seconds was your rocket in the air?*

*Overall, are you happy with your design?*

*Would you change your design if you could? How do you predict such a change would affect the rocket's motion?*

• When all groups have presented, return to the classroom and distribute a Challenge Reflection sheet for each student to complete.

## *Opportunities for Differentiation*

**To make it simpler:** Choose one option for powering the rocket. Provide a basic rocket design for students to begin with and allow students to modify the basic design. Provide paper towel or bath tissue tubes to use as body tubes.

**To make it harder:** Provide fewer of each material in each kit. Require a specific propellant, such as seltzer tablet and water or baking soda and vinegar. Include a minimum distance and height for the rocket's flight.

Names _____    _____

_____    _____

# Rocket Test Results

| Criterion | Score |
|---|---|

## 1. Length

Length between 6 and 12 inches, score = 1

Length less than 6 inches or more than 12 inches, score = 0

    _____

## 2. Weight

Weighs 1 pound or less, score = 1

Weighs more than 1 pound, score = 0

    _____

## 3. Propellant

Propelled by force not directly from a human, score = 1

Propelled by direct human force, score = 0

    _____

## 4. Horizontal distance traveled

Score 1 point for each foot traveled (round to the nearest foot)

    _____

## 5. Height reached

High, more than 7 feet, score = 3

Medium, 4 to 7 feet, score = 2

Low, less than 4 feet, score = 1

No flight, score = 0

    _____

## 6. Travel time

Score 1 point for each second under 10 seconds

Subtract 1 point for each second over 15 seconds

    _____

## Total Score

    _____

# Catapults

## Curriculum Connection

Science: forces & motion, potential (stored) energy

## Criteria for Product

- Lighter, easier-to-carry catapults score higher than awkward, heavier ones
- Freestanding catapults score higher than catapults that must be held down
- Catapults should send the marshmallow at least 6 feet, with a higher score for longer distances

## Constraints for Challenge

- Must use only the materials provided
- Must complete each stage in the time allowed
- Only three attempts are allowed, with the longest distance used for scoring

### Challenge

From ancient times through current day, humans have launched objects through the air. Students design and build catapults to launch safe projectiles—marshmallows.

## Materials

### Catapult Materials (for the class)

- 25 plastic spoons
- 100 jumbo-size craft sticks
- 100 regular-size craft sticks
- 100 assorted rubber bands
- 25 assorted sizes of binder clips
- 50 paper plates
- Several sticks, either dowels or from outdoors (assorted sizes and lengths)
- 12+ egg cartons
- 50 straws
- 100 clothespins (with springs)
- 100 unsharpened pencils
- 30 bottle caps, any size or style
- 50 small paper cups, 6-oz. size
- 100 paper clips, any style or size
- 2 rolls of clear tape
- 2 rolls of masking tape
- Assorted shapes and sizes of cardboard boxes

### Tools

- Scissors
- Hole punches

### Additional Materials

- Chart paper
- Markers
- Plain paper
- Pencils
- Digital timer
- Wood block
- Rock, large enough for class to see
- Metal spoon
- Yardstick or tape measure
- Sidewalk chalk, cones, or other means of marking distance
- 5 bags of regular-size marshmallows
- Digital video camera (optional)

# Before You Begin

- Scout out a good testing place for the marshmallow catapults—outdoors is ideal. Look for a place with no obstacles, such as trees or playground equipment. If testing must be indoors, look for an empty hallway, gym, or cafeteria. Review the set-up instructions under Test & Present for more information.

- Also designate one or more areas within the classroom where students can try out their catapults while they build. A clear area on the floor next to a wall is a good option.

- Review with students the engineering design process and related definitions for engineering, criteria, constraints, and brainstorm.

- Make a set of Job Cards (page 13) and a copy of the Test Results (page 29) for each group.

- Make a copy of the Challenge Reflection sheet (page 14) for each student.

- Set the catapult materials out on a table.

- Set up the digital timer where students can check on it.

- Display the Criteria and Constraints where students can see them.

# 5-Step Process

**1** *Investigate*
*Session 1*
*30 minutes*

- Ask students to share anything they know about energy.

- Explain to students that energy that is available for use is called stored energy or potential energy. Ask students for examples of things that have stored energy, such as gasoline, batteries, and food. Tell students that in this Challenge, the focus is on storing and using one specific kind of energy, *elastic stored energy*, or *elastic energy* for short.

- Invite students to offer definitions of elastic. Bring the discussion around to recognizing that elastic materials can bend or be squeezed and then snap back to their original shape. Then place a rubber band, plastic spoon, clothespin, craft stick, binder clip, wood block, metal spoon, and rock in front of the students. Call one student at a time to come up to the table, choose an item, and explain to the class whether or not it has elastic properties. Instruct students to demonstrate how the object is or is not elastic.

- After all items have been discussed, ask students to sort the items into categories that describe elasticity. Allow students to come up with the categories. Examples of categories include "completely elastic," "parts that are elastic," "elastic up to a point," or "not elastic." For example, the rubber band might be "completely elastic," while the clothespin has "parts that are elastic." The craft stick and plastic spoon might be "elastic up to a point" because they break if bent too far. The rock and block are "not elastic."

**1** *Investigate*
*Session 2*
*20 minutes*

- Remind students about the work from the last session and review the concept of elastic energy.
- Let students know that throughout history, engineers have used elastic energy to build machines to send items flying through the air. These machines are called catapults and the objects they send flying are called projectiles. Catapults have been used as weapons, to move heavy items from one place to another, and to release airplanes from ocean-going aircraft carriers.
- Tell students that for this Challenge, they will use their knowledge of elastic energy to design and build catapults that launch marshmallow projectiles. Ask students:

  *What properties of materials do you need to consider when you design?*

  *How does a catapult get energy to store? What can you do to it to give it energy?*

  *What are some different ways that you can store elastic energy?*

  *Can a catapult be completely elastic, or should parts of it be stiff (not elastic)?*

  *What questions do you have about the Challenge?*

- Review with students the Criteria and Constraints. Display the Test Results sheet so students know exactly how their catapult will be judged against the Criteria. Also review the time limits for Brainstorm, Plan, and Build.

**2** *Brainstorm*
*30 minutes*

- Show students the items on the materials table. Tell them that they can use any of these materials—but no others—to build their catapults.
- Invite a few students at a time to come examine the materials. Ask students:

  *What properties do each of these materials have?*

  *How could the properties of the materials help a catapult to work?*

- Review the meaning of brainstorm. Tell students that before they begin building, they first need to come up with ideas and then make a plan.
- Give each student a sheet of plain paper. Have students fold the paper in half vertically and again horizontally, creating four sections when unfolded. Then have them write Brainstorm along one edge.
- Ask students to work alone quietly for 15 minutes. Their task is to think of four different ideas for building a catapult, using the materials they just looked at. Students should draw a different idea in each of the four boxes. If a student has more than four ideas, suggest using the other side of the paper.
- As students work, walk around the room and ask guiding questions to extend student thinking. Suggest starting places for those who are stuck. Sample questions:

  *Why did you choose that material instead of another?*

  *How does your catapult work?*

*What part of your catapult holds the marshmallow?*

*Which design is your favorite? Why?*

- After 15 minutes, ask each student to choose one favorite idea and circle it. Explain that they will get into groups. Each group will come up with one design to build. It might be like their favorite design, or it might be a different design.

## 3 Plan

*30 minutes*

- Think of your class in terms of four levels of academic ability. Then create mixed-ability groups of three to four students each. Have students sit with their groups.

- Explain that all group members help plan and build the catapult. But each group member also has one of four special jobs.

  - **Speakers** make sure everyone gets a chance to talk. They also bring questions to the teacher, help solve disagreements before asking the teacher to help, and answer questions when the group presents.

  - **Timekeepers** keep track of the time, let everyone know how much time is left, and help keep everyone on task. They may assign tasks as needed.

  - **Materials Managers** get materials for the group and test the catapult.

  - **Recorders** write down or draw all of the group's work, including the plan and the test results.

- Place a set of four Job Cards facedown next to each group. Have each student choose one job card at random. (For groups with three students, one student takes two job cards.) Tell students that they may not trade jobs or groups, but they will do different jobs and be in different groups during other Challenges.

- Set the timer for 20 minutes for groups to Plan. Have students explain their ideas to each other and then decide on one idea to build. Remind students that each person needs to share before the group decides. Make sure students understand that they may also combine ideas or change an idea to make it better.

- Circulate around the room to offer support as needed and to mediate disagreements. Sample questions:

  *Did you choose one person's idea or combine several ideas?*

  *How can you disagree with someone and still be respectful?*

  *How are you all showing that you can listen actively to each other?*

  *Has everyone had a chance to share their ideas?*

  *How did you go about sharing ideas? Did you talk about one idea at a time or more than one idea? Did you talk about ways to make an idea work better?*

- After a group agrees on one design, provide a sheet of chart paper. Ask the Recorder to draw and label a diagram of the group's catapult. The plan needs to include a list of materials and an estimate of how much they'll need of each material.

## Build

**45 minutes**

- Tell students it's time to build. While they are in the classroom, they may launch their catapults only when they are facing the wall in the designated testing area. Otherwise, things or people could be hurt.

- Set the timer for 45 minutes. Have the Materials Managers collect the materials that the group listed.

- Remind students to begin by following their plans. A group may change their plans if they want, but first, the whole group must agree to make the change. Then, the Speaker must get permission from the teacher. The Recorder then uses a different color marker to record any changes to the plan. This process takes time, so the Timekeeper needs to advise the group if there is time to make the change.

- As students work, circulate around the room. Ask guiding questions as appropriate, but encourage students to solve problems on their own in their groups. Sample questions:

  *What part of the build is each person working on?*

  *Is the catapult coming out the way you imagined?*

  *What do you know about stored energy that you are thinking about as you build?*

  *Is your catapult easy to move from place to place?*

  *Have you done a test yet?*

  *Did you change your catapult after a test? If so, why?*

- Store the finished catapults in a safe place until it is time for Test & Present.

## Test & Present

**3–5 minutes per group**

- Bring students to the testing area with their catapults, a yardstick or tape measure, and chalk or objects to mark the distances. Bring the digital video camera if you plan to record each presentation.

- Have student volunteers mark a starting line. Assist volunteers in using the yardstick or tape measure to measure and mark distances in 2-foot increments up to 10 feet. (If it looks like your students' catapults will have a different range, adjust the increments to suit their results.)

- Invite the first group to present their catapult to the class. First ask questions as noted below. Then give the Recorder a Test Results sheet and a pencil. Have the Materials Manager place the catapult on the starting line, position the marshmallow, and shoot. Ask the Recorder to check the distance the marshmallow traveled, report it, and mark it on the Test Results sheet.

- Ask questions before, during, and after each demonstration. Sample questions:

  **Before the demonstration:**

  *How does your catapult work?*

  *Is your catapult easy to move from place to place?*

  *Does your catapult stand up on its own?*

  *Did you follow your original plan or did you make changes?*

*Why did you make those changes?*

*Give one example of how your group cooperated.*

**During the demonstration:**

*Is your catapult working the way you planned?*

*How far did your marshmallow fly on the first attempt?*

*Is the catapult holding up well during the test?*

**After the demonstration:**

*Do you think your design was successful? Why?*

*What detail of your catapult do you think made the marshmallow fly as far as it did?*

*What part of your design could you improve?*

- After all groups have presented and all materials have been brought back indoors, have each student complete a Challenge Reflection sheet.

- If you made videos, arrange for a showing in class. Use the opportunity to compare the catapults and to discuss what parts of each design contributed to the results for that catapult.

## *Opportunities for Differentiation*

**To make it simpler:** Show students how to use the plastic spoon as the basis for the catapult by placing the marshmallow on the bowl of the spoon, pulling it back, and letting it go. Instruct groups to use the materials provided to improve this catapult so that it meets all of the Criteria.

**To make it harder:** Remove the plastic spoons and cardboard boxes from the available materials. Tell students they can use only a certain number of materials in their design.

Names _____    _____

_____    _____

# Catapult Test Results

## Criterion                                                      Score

### 1. Portability                                               _____
Easy to carry and lightweight, score = 4
Easy to carry but heavy, score = 3
Awkward to carry but lightweight, score = 2
Awkward to carry and heavy, score = 1

### 2. Ability to stand alone                                   _____
Stands on its own, score = 3
Must be attached to something else to stand up, score = 2
Falls down when nothing is holding it, score = 1

### 3. Longest distance marshmallow traveled                    _____
Distance on attempts 1, 2, and 3; use the longest distance for scoring.

     1. _____ feet

     2. _____ feet

     3. _____ feet

Score 1 point for every 2 feet traveled up to 6 feet
(round to the nearest 2 feet)
Score 1 additional point for each foot over 6 feet traveled

### Total Score                                                 _____

# Compasses

## Curriculum Connections

Science: magnetism
Social Studies: explorers

## Criteria for Product

Compass will be scored on:

- Pointing north; higher scores for working in more than one location
- A case that encloses it; sealed cases score higher
- Visibility while sealed in case
- Labels: north, east, south, and west labeled correctly
- Portability and ease of use: ability to use with one hand scores highest

## Constraints for Challenge

- Must use only the materials provided in the group kit
- Must complete each stage in the time allowed

### Challenge
Today we have global positioning satellite systems, but for centuries sailors relied on simple magnetic compasses to find their way at sea. Students design and build magnetic compasses.

## Materials

### Compass Materials (for each group)

- 1 steel needle with blunt tip
- 1 strong bar magnet
- 1 bottle of water, any size
- 1 plastic bottle cap (in addition to cap on water bottle)
- 4 clear plastic cups, 12-oz. size or larger
- 3 small paper drink cups, 6-oz. size
- 1 cork
- 1 foam tray
- 1 aluminum pie pan
- 2 straws
- 5 uncoated metal paper clips, any size
- 6 craft sticks
- 1 sheet of plain paper
- 1 roll of clear tape
- 1 sq. ft. of plastic wrap
- 1 ft. of thread
- 1 shopping bag (to hold materials kit)

### Tools

- Hole punches
- Scissors
- Permanent markers

### Additional Materials

- A few dozen assorted magnets
- A few additional uncoated metal paper clips
- Permanent marker
- Colored markers
- Chart paper
- Plain paper
- Pencils
- Digital timer
- Masking tape
- Digital video cameras (optional)
- Method of displaying videos (optional)

# Before You Begin

- Assemble a materials kit for each group, following the Compass Materials list.
- Set out magnets and paper clips for Investigate, Session 1.
- Choose three different locations, indoors or outdoors, to test the compasses. Use your classroom as one location if possible. The locations need to be well away from steel doors or sources of magnetic fields, such as generators. Determine which direction is north at each of the three locations and note this information.
- Make a set of Job Cards (page 13) and a copy of the Test Results (page 36) for each group.
- Make a copy of the Challenge Reflection sheet (page 14) for each student.
- Set up the digital timer where students can check on it.
- Display the Criteria and Constraints where students can see them.

# 5-Step Process

 **Investigate**
**Session 1**
*30 minutes*

- Have students form pairs. Provide each pair with a variety of magnets and paper clips. Give students about 10 minutes to freely explore these materials.
- After free exploration, start a class discussion about the properties of magnets. Suggested questions:

  *What happens when you place two magnets near each other?*

  *How can you use one magnet to figure out the poles of another magnet? (Hint: It helps if one magnet has labels on its north and south poles.)*

  *What happens to a magnet when it can turn freely and is not near another magnet?*

- Guide the class discussion to a review of the properties of magnets. Students should remember that like poles repel each other and opposite poles attract each other. Remind students that Earth itself is a magnet, and that magnetized needles turn to line up with Earth's magnetic field when they are allowed to.
- Show students how to magnetize a blunt needle or paper clip, using a strong magnet. Hold the needle or paper clip in one hand and the magnet in the other. Stroke the needle or paper clip with the magnet at least 10 times, always stroking in the same direction, not rubbing back and forth.
- When the needle or paper clip has been magnetized, tie a thread to the center of it so it hangs horizontally and swings freely. When it stops swinging, tell students that it has lined up with Earth's magnetic field. Ask them which end points north. (Students can reasonably guess either end of the needle, so after letting them guess, tell them which direction is north.) Use a permanent marker to darken the end of the needle or paper clip that points to Earth's North Pole. (Note: Technically, the north end of

a magnet is the north-seeking end of the magnet because it is attracted to Earth's North Magnetic Pole. However, it's simpler to call this the north pole of the magnet, and that is what it's called in this book.)

**1** *Investigate*
*Session 2*
**30 minutes**

- Remind students how you were able to magnetize the paper clip or needle. Explain that this metal item now has north and south magnetic poles. Tell students that the same end of this item will always point north if it's allowed to turn freely.

- Explain to students that when Christopher Columbus wanted to cross the ocean, he needed to know which direction to go, so he used a device called a compass. Compasses are a very old technology, but like most technology, they take a lot of planning and testing to build. If students haven't already guessed, tell them that they'll be designing and building compasses.

- Explain to students that early compasses were made from either a sliver of magnetite, a naturally magnetic mineral, or from a magnetized metal needle. The magnetic material was placed on wood or hay and floated in water. Since the needle was floating, it was able to turn freely so it could point north. Tell students that compasses today are based on this same idea, but they are smaller and more advanced.

- Read the Criteria and Constraints and explain anything that isn't clear. Show students the Test Results sheet so they know how their compasses will be scored. Point out the time limits for Brainstorm, Plan, and Build.

**2** *Brainstorm*
**20 minutes**

- Show students the materials kit. Tell them they may use any materials in the kit—but no other items—to build their compasses.

- Set out samples of each material found in the kits so that students may touch and observe them while brainstorming.

- Give each student a sheet of plain paper. Have them write Brainstorm at the top. Instruct them to work silently and independently for 15 minutes. Tell them to think of as many different designs for a compass as possible. Explain that they may include different materials from the kit for each design, or they may use the same materials in different ways.

- Encourage students to label diagrams with the materials they plan to use, for example, needle, plastic cup, thread.

- While students brainstorm, walk around the room. Ask guiding questions to encourage detailed work and offer starting places for students who are stuck. Suggested questions:

  *How will you set up the magnetic pointer so that it moves freely?*

  *What are you planning to use as the base?*

  *What properties of magnets do you need to think about when you choose materials?*

  *Which materials would work best for a see-through case to hold the compass?*

  *What questions do you have about the Challenge?*

- After 15 minutes, ask students to choose one favorite idea and circle it, and to make sure their names are on their papers. Collect all Brainstorm papers.

**3** **Plan**

*30 minutes*

- Before Plan begins, review the "favorite idea" designs on all Brainstorm papers, and then make groups of three to four students each by placing students with similar designs together. For example, group together those who plan to use the same material for the compass's case.

- Return the Brainstorm papers to students by group. Give each group a set of Job Cards. Instruct each group to turn the cards facedown on the table, and then have each student choose one card at random. The student does the job named on the card: Speaker, Materials Manager, Timekeeper, or Recorder. (For groups of three, one student chooses two job cards.) Students may not trade jobs. Review the responsibilities of the four jobs (page 12). Remind students that each group member helps to plan and build, as well as do a group job.

- Set the timer for 20 minutes. Instruct students to review their favorite designs with others in their group, looking for similarities. Encourage groups to use similarities as a starting place for a group plan. Remind them that after all members have shared, the group comes up with one plan that they all think will work. Encourage groups to use additional pieces of paper to jot down ideas as they talk.

- As students discuss, circulate around the room to offer support as needed and mediate disagreements that the Speakers cannot manage. Sample questions to ask students as they work:

  *How are your four favorite designs alike?*

  *How is your final design like your four favorite designs? How is it different?*

  *Did you choose one person's design or did you use parts of two or more designs?*

  *Which materials are you planning to use from the kit? How will you use them?*

  *How do you plan to hold up the pointer? How will it be able to turn?*

- After each group agrees on one plan, the Recorder leads the group in writing the plan on chart paper using colored markers. The plan needs a diagram and labels. Groups should draw each material in a different color and include a key, so that the diagram is easy to understand (for example, plastic wrap drawn in blue, craft sticks drawn in brown).

### Build
**60 minutes**

- Tell students they will work as a group to build and test their compasses. Set the timer for 60 minutes and invite each Materials Manager to pick up a materials kit.

- Remind Timekeepers to watch how much time is left, make sure the group is making progress, and participate in the construction of the compass. As groups work, check in with the students doing the other jobs, as well.

- Encourage students to test their compasses from time to time and then make changes as needed. Remind students that if a group agrees to make a change, the Speaker must first notify the teacher. Then the Recorder makes changes to the plan in a different color or draws up a Plan 2.

- As students work, monitor the progress of each group. Remind students to use teamwork and cooperation. Sample questions to ask while monitoring:

  *What part of the compass are you building now?*

  *How are you settling disagreements?*

  *Is the design coming out the way you thought it would?*

  *What do you know about magnets and magnetic poles that you are considering as you build?*

  *Why did you choose that material?*

  *Have you tested to see if your compass works? What did you learn?*

- When a compass is ready for its final test, the Speaker should notify the teacher. If you are using videos, then have groups proceed to Test & Present as they finish.

### Test & Present
**5–10 minutes per group**

- If you are having groups make videos of the tests, then let each group borrow a video camera as they finish, and have the Recorder use it to record the test, following the procedure below. If groups are presenting tests live, then set aside the compasses until it is time to test, and bring all groups to the three test locations as a class.

- The testing procedure applies whether students are presenting live or recording the tests on video. Students visit each of the three locations you chose to test item 1 on the Test Results sheet. Be sure to tell students which direction is north in each location. During each test:

  - The **Recorder** captures the test using a video camera (optional) and completes the Test Results sheet.

  - The **Materials Manager** holds the compass so that the needle can turn.

  - The **Speaker** answers questions and narrates the test, explaining what is happening.

  - The **Timekeeper** looks to see which way the compass is pointing.

- Time permitting, you may allow students to work on an improved compass design after the test.

- Encourage the class to ask questions, and have the Speaker answer them either during live presentations or while showing videos. Sample questions:

**Before the demonstration:**

*Which materials did you use?*

*How were your group's favorite Brainstorm plans alike?*

*How many times did you change your group's original plan?*

*Who came up with an idea that no one expected?*

**During the demonstration:**

*Does the compass work the way you expected?*

*Why is the compass acting this way?*

*How is the case holding everything together?*

*Can you see the needle clearly through the case?*

**After the demonstration:**

*Overall, was your compass design successful? How do you know?*

*Which part of your design would you improve?*

*What other items could you design that use magnets?*

- When all groups have presented, return to the classroom and instruct each group to score their compasses for items 2, 3, 4, and 5 on the Test Results sheet.
- Distribute a Challenge Reflection sheet to each student.

## Opportunities for Differentiation

**To make it simpler:** Show students how to float a magnetized needle on some hay in a shallow dish of water. Once the needle stops spinning, remind students that the north pole of the needle is pointing north, and have them suggest a way to label that end of the needle. After this demonstration, tell students to make a smaller, more portable version of this compass.

**To make it harder:** Remove the bottle of water from the kits so students must think of another way to suspend the pointer inside a case.

Names _____  _____

_____  _____

# Compass Test Results

## Criterion                                              Score

### 1. Compass points north                              _____

In three locations, score = 3
In two locations, score = 2
In one location, score = 1
Does not point north, score = 0

### 2. Compass case                                      _____

Completely closed, score = 2
Closed, but comes loose easily, score = 1
Not closed, score = 0

### 3. Able to see needle inside case                    _____

Can read the compass easily, score = 2
Can read the compass with some difficulty, score = 1
Cannot see inside to read the compass, score = 0

### 4. Compass directions labeled                        _____

Four directions are labeled correctly, score = 4
Three directions are labeled correctly, score = 3
Two directions are labeled correctly, score = 2
One direction is labeled correctly, score = 1
No directions are labeled correctly, score = 0

### 5. Portable and easy to use                          _____

Portable and easy to use with one hand, score = 3
Portable and easy to use with two hands, score = 2
Can be moved to a level surface for use, score = 1
Cannot be moved, score = 0

### Total Score                                          _____

# Healthy Snacks

## Curriculum Connections

Mathematics: measurement, fractions
Health: nutrition

## Criteria for Product

- Should be tasty and have a good texture
- Must be easy to carry and eat on the go
- Flavors should blend well together
- Should look good to eat
- Must be nutritious
- Must contain at least four ingredients

## Constraints for Challenge

- Must use only the materials provided
- Must complete each stage in the time allowed
- Must not need refrigerated storage
- Must not need a heating device to prepare (such as a stove or microwave)

> ### Challenge
> It can be difficult to eat well when you're in a hurry. Students design and make nutritious, healthy snacks that can be eaten "on the go."

## Materials

### Healthy Snack Ingredients (for the class)

- To be determined during class Brainstorm
- Alternatively, start with the Ideas list (page 45)

### Tools

- Plastic spoons and forks
- Blunt table knives or sturdy plastic knives
- Measuring spoons and cups
- Mixing bowls
- Mixing spoons
- Pastry blenders
- Potato mashers
- Foil baking pans (6 × 9 in. work well)
- 1 roll of aluminum foil
- 1 roll of parchment paper
- 1 roll of clear plastic wrap

### Additional Materials

- Packaging from common snacks, with nutrition information
- Computer with Internet access (optional)
- Projector for computer (optional)
- Plastic food-service gloves
- Access to a refrigerator (optional)
- Chart paper
- Markers
- Plain paper
- Pencils
- Hula hoops or long lengths of yarn or rope
- Digital timer
- Ingredients for practice snack bar (recipe page 45)

**Note:** *Substitute suitable alternatives for ingredients that pose an allergy risk for any of your students.*

# Before You Begin

- Collect snack packaging for Investigate—for example, candy-bar wrappers, chip bags, cookie bags, donut boxes, fruit-flavored snacks, and so forth.

- Download the MyPlate Mini-Poster to show to students. Posters can be downloaded from the USDA website at (www.choosemyplate.gov); look for links to printable materials at the bottom of the webpage.

- While at the USDA websites, preview and bookmark the Food Groups Overview webpage (www.choosemyplate.gov/food-groups) or look for links to printable MyPlate resources about food groups.

- Make two large labels by writing Healthy on one sheet of paper and Not Healthy on another sheet, using a wide marker.

- Decide whether you want to brainstorm a class ingredients list, or if you prefer to shop in advance from the Ideas list (page 45).

- Plan to review the ingredients list and make any substitutions needed due to food allergy concerns or cost. Then, either contact families for donations of chosen ingredients or go shopping.

- Make a set of Job Cards (page 13) and a copy of the Healthy Snack Survey (page 46) for each group.

- Make a copy of the Challenge Reflection sheet (page 14) for each student.

- Arrange to use a refrigerator for the Investigate and Build stages.

- Write the numbers 1, 2, 3, and 4 on separate slips of paper so that each group has a set of numbers 1 to 4.

- Set up the digital timer where students can check on it.

- Display the Criteria and Constraints where students can see them.

# 5-Step Process

**1 Investigate**
*Session 1*
*45 minutes*

- Ask students what they know about how the human body gets its energy. Give several students the chance to share their thoughts. As needed, guide students to understand that humans get energy from food. Without proper nutrition, we humans become tired and weak.

- Display the MyPlate Mini-Poster. Point out that foods are organized into different groups. Each part of the MyPlate diagram shows a different food group. It is important to eat foods from each group daily, and the size of the space on MyPlate shows how much of our daily food should come from each group. Show the lists of foods in each group from the MyPlate website, and discuss foods that fall into each category.

- Point out that many snack foods don't belong in any part of the MyPlate diagram. A food that does not have a place on MyPlate does not make you healthy. Foods that are not healthy for you tend to have a lot of sugar, a lot of fat, or a lot of both.

- Place two hula hoops on a clear area of floor, slightly overlapping them to create a Venn diagram. Alternatively, you could use two long pieces of yarn or rope. Using the labels you made, title the left circle Healthy and the right circle Not Healthy. Place the snack packages outside the diagram. Gather students around the Venn diagram.

- Invite students, one at a time, to choose a snack package and place it in the Venn diagram. Students should explain how they decided to place each package. If a student is unsure of where to place a package, suggest she read the nutrition label for sugar and fat content.

- When all the packages have been placed, ask students what they can conclude about most of these snacks, based on the diagram. Invite students to think about how many of the snacks they just sorted are items that they enjoy. Tell them to think about the properties of a "good snack"—that is, a snack that many people like to eat. Ask students:

  *Why do you choose to eat these snacks? What is it that you like about them?*

  *What do you mean when you say, "They taste good"? Be more specific.*

  *What are the most common characteristics you look for in a snack?*

  *Is it important for a snack to be easy to eat away from home?*

  *What makes a snack easy to eat in the car, at school, or at sports events?*

  *What kinds of snacks stay fresh away from home? What kinds do not?*

  *Are loose snacks like cereal, nuts, and granola easy to eat while you're moving around?*

  *Do you prefer snacks that you don't need a spoon to eat?*

- If students do not offer the following suggestions, guide them to descriptive words such as *sweet*, *salty*, *crunchy*, and so forth. Students will probably conclude that good snack foods can be held in one hand, do not require utensils to eat, can be taken anywhere, and stay fresh without refrigeration—and taste good, of course.

- Write down the properties of a "good snack" on chart paper for later use.

❶ *Investigate Session 2*

*45 minutes, then 10 minutes to sample snack*

- Tell students that there is a huge push in the United States and elsewhere to help kids eat healthier foods, including snacks. But many healthy snacks have to be eaten sitting down, or with a spoon or fork. Many require refrigeration or cooking.

- Food companies hire engineers to find ways to process raw ingredients into snack foods. The foods they produce have everything we love about snacks, and they suit the way we like to live and eat. Unfortunately, many of the resulting snacks are unhealthy.

- Explain to students that their next Challenge is to design and make snack bars that are healthy, yummy, and easy to eat "on the go." To prepare for this Challenge, students need to think about liquids, solids, and textures of different foods, and how the combination of two or more ingredients can change the properties of both. Ask students:

  *Have you ever cooked or baked with your parents?*

  *What happens when you mix ingredients together?*

  *Are certain ingredients better for "sticking things together" than others?*

  *How can you make a liquid thicker?*

  *How can you make something that is dry, stickier?*

- Show students the ingredients for your trial snack bar. Ask students:

  *Which ingredient might make the other ingredients stick together?*

  *What could you do to combine the ingredients as thoroughly as possible (for example, stir, mix, roll, mash, press, crunch, crumble)?*

  *What could you do to make soft items firmer or more solid?*

  *How could you make sticky things drier?*

- Tell students that you have a recipe (page 45) for a snack. A recipe is the food engineer's plan. Like all engineering plans, a recipe describes a process to follow in order to get the best product. It is important to follow the process exactly, using the correct amount of each ingredient, following the suggested order, and doing the actions specified (stirring, rolling, and so forth). If you do not follow the process, then the results may not be what you intended. When food engineers design a new snack, they have to come up with the recipe. Tell students they'll be doing that for their Challenge. For now, they'll watch how one snack bar is made.

- Display the snack bar recipe where students can see it. Wash your hands, put on food-service gloves, and then follow the recipe step by step, exactly as it is written. Explain that this recipe makes six full-size (2 × 3 in.) bars. You'll divide it into smaller bite-size pieces so everyone can taste-test it. As you work through the recipe, ask students:

  *What might happen if I added more of this ingredient? What if I put in less?*

  *What properties does this ingredient have? How does it help the snack meet the "good snack" criteria?*

  *Why do I need to combine these ingredients separately from the others?*

  *How does it help to have some of the cereal mashed in and some rolled on later?*

  *How could I adjust the recipe to make enough for 100 people? Would I be able to mix it the same way?*

  *After pressing the mixture into a pan, it helps to put it in the refrigerator or another cool place. Why do you think that helps?*

- Put the bars in the refrigerator to harden. At least 2 hours later, cut the bars into enough pieces so everyone can sample them. Invite everyone to offer their thoughts and ideas. Ask students:

  *Do all of the flavors seem to blend nicely?*

  *What is the texture of the snack bar? Do you like it, or do you think it is too dry, too soft, too hard, or too chewy?*

  *Is the snack bar easy to eat?*

  *What might you change about the recipe? Why?*

  *Does everyone have to agree about liking the snack bar?*

- Review with students the Criteria and Constraints, as well as the time limits for Brainstorm, Plan, and Build. Ask if they have any questions about the Challenge. Share the recipe with your students' families, in case students wish to make the bars at home.

**2** **Brainstorm**
**30 minutes**

- As a class, take a look at the list of "good snack" properties from Investigate. Ask students to think of ingredients that could go into a snack that is healthy but still has many of the properties on the list. Ask students:

  *What ingredients are creamy…crunchy…chewy…salty…?*

  *What ingredients are naturally sweet?*

  *What ingredients can help change the properties of other ingredients to make them easier to work with (for example, to make them drier, thicker, or softer)?*

  *Should you include any ingredients that need to be refrigerated? Why or why not?*

  *Are there any ingredients that you can substitute for those that are unhealthy?*

- An alternative approach for Brainstorm is to gather ingredients from the Ideas list (page 45) and display them before beginning Brainstorm. Using this approach, discuss as a class different ways that the ingredients could be combined to produce a good snack bar.

**3** **Plan**
**30 minutes**

- Consider the learning styles among your students—for example, kinesthetic, visual, or auditory. Make groups of three to four students so that each group has different learning styles mixed together. Do not consider academic levels for this Challenge.

- Once students are grouped, provide each group with a set of paper slips numbered 1 to 4 and a set of Job Cards. Have each student within each group choose a numbered slip at random and then choose Job Cards in that order, 1 to 4. (In groups of three, the student who draws first takes two jobs.) Review the responsibilities of the four jobs, Speaker, Materials Manager, Timekeeper, and Recorder (page 12). Remind students that they may not trade jobs, but they will do a different job on another Challenge.

- Tell students that for this Challenge, Plan involves coming up with a recipe for a snack bar. The recipe must include an ingredients list, measurements for each ingredient, a list of kitchen tools needed, step-by-step instructions for assembling the snack bar, and a drawing of the finished snack. Their recipe needs to list enough of each ingredient to make six full-size snack bars, but they'll cut the finished snack into smaller pieces for taste-testing. Finally, they should come up with a name for their snack bar.

- Explain that when Plan time begins, groups need to look at the ingredients list from Brainstorm. In the same 1 to 4 order, each member of the group will take a turn naming favorite ingredients from the list and suggesting how to combine them. After all members have had a turn, the group comes up with one recipe that they all agree to follow. Answer any questions students may have about what to do, and then set the timer for 20 minutes.

- As students share ideas, circulate to offer support and mediate any disagreements that the Speakers cannot resolve. Sample questions to ask students:

  *How does your recipe show that you understand the properties of a good snack?*

  *Is anything in your recipe not part of the Criteria? Why did you include it?*

  *Is your snack creamy, chewy, or crunchy?*

  *Explain how each person contributed to the recipe.*

  *What is it about this plan that you all like?*

  *How much of each ingredient will you use? How did you decide on that amount?*

  *How do you know that your snack will make enough for everyone?*

- As each group agrees on a recipe, provide chart paper and markers. The Recorder assigns each person in the group part of the plan to write or draw on the chart paper. The ingredients list needs at least four ingredients and cannot have any ingredient that is not on the table.

**Build**

*60 minutes*

- Tell students they will work as a group to make their Healthy Snacks. They need to be sure their recipe is large enough to make six servings. Set the timer for 60 minutes. Invite Materials Managers to gather enough ingredients to make a sample of the snack for each student, as well as any tools they need. Be sure each group takes a pan.

- Make sure all students wash their hands before beginning. Give each student a pair of plastic food-service gloves to wear while preparing the food. Be sure students understand that they may touch nothing but food and kitchen tools while wearing the gloves. They must wash their hands and change gloves if they touch something else.

- Remind students to follow their recipe. They may change the recipe if it looks like they need to, but first all students in the group must agree. Then the Speaker gets permission from the teacher and the Recorder makes changes to the plan using a different color.

- As students work, walk around the room, asking questions and encouraging students to solve problems on their own. Sample questions:

  *How is each person helping?*

  *Is your group using good teamwork skills?*

  *Are all the ingredients working the way you planned?*

  *Have you made any changes to your plans?*

  *Are you measuring as you go to be sure you are following your recipe?*

- Remind students that everyone helps "build" in addition to doing a group job.

- Place finished snack pans in the refrigerator. Allow snack bars to chill and harden for at least 2 hours before Test & Present.

**Test & Present**

*5–10 minutes per group*

- Tell students that when food engineers come up with new foods, they do taste tests with large groups of people. They use the results of the taste tests to make changes and improvements to their recipes. Explain that each group will conduct a taste test with the class and complete a survey of the results to get feedback for improvement.

- Have all the Materials Managers and Timekeepers wash their hands. Give each Materials Manager a sturdy plastic knife. Have them work with the Timekeepers to cut their snack so that everyone in the class can have a small piece.

- Call the first group up to the front of the room to present their snack. Hand the Recorder a copy of the Healthy Snack Survey.

  - The **Speaker** tells the group about the snack and answers questions about it.

  - The **Materials Manager** and **Timekeeper** hand out samples of the snack.

  - The **Recorder** calls for a show of hands for "Definitely no," "Sort of," and "Definitely yes" for each question in the survey, counts up the number of votes, and records them on the Survey sheet.

- Remind groups to try their own snacks and include their own votes in the survey.

- Encourage the class to ask questions and have the Speaker answer them during each group's turn. Sample questions in case students need prompting:

**Before the test:**

*What do you call your snack?*

*What was one challenge your group faced?*

*What different properties of good snacks did you include?*

**After the test and survey:**

*Overall, was your recipe successful? How do you know?*

*What part of your recipe would you improve?*

*Would your recipe work if it had to be made ahead of time and packaged?*

• When all groups have presented, ask Recorders to work with the rest of their groups to tally up the final score on the survey sheet.

• Have each student complete a Challenge Reflection sheet.

## *Opportunities for Differentiation*

**To make it simpler:** Rather than having students come up with their own recipes, give them copies of the recipe on page 45. Instruct them to work from that recipe but substitute different ingredients, leaving the procedure and amounts the same.

**To make it harder:** Challenge students to investigate the nutritional information for each of the ingredients they've included in their recipes. They should multiply the nutrient information for that ingredient by the amount of each ingredient in the recipe, and then divide it out again to a "per serving" figure, assuming that the recipe makes six servings.

# Ideas for Healthy Snack Ingredients
### For reference during Brainstorm, or to use as a shopping list

### Crunchy Foods
- Popcorn
- Granola
- Pretzels
- Rolled oats
- Cereal (low sugar)
- Rice cakes
- Wheat germ
- Nuts
- Sunflower seeds
- Pumpkin seeds
- Sesame seeds
- Dark chocolate
- Coconut

### Sticky Foods
- Dried fruits
- Peanut butter
- Almond butter
- Applesauce
- Honey
- Molasses

### Thickeners
- Nonfat dry milk
- Flour
- Gelatin dessert mix
- Pudding mix
- Cocoa powder

### • Flavorings
- Salt
- Nutmeg
- Cinnamon
- Allspice
- Ginger
- Vanilla extract

---

# Apple Cinnamon Snack Bars Recipe
### (makes six servings)

| | |
|---|---|
| ½ c. applesauce | 1½ c. rolled oats or quick oats |
| 1 tbsp. honey | ¾ c. toasted rice cereal |
| ¼ tsp. vanilla | ¼ tsp. salt |
| 2 c. nonfat dry milk | 1½ tbsp. cinnamon |

1. In large mixing bowl, combine applesauce, honey, and vanilla.
2. Add salt and cinnamon. Mix well.
3. Gradually stir in dry milk.
4. Fold in oats, followed by rice cereal.
5. Check that all dry ingredients are well-coated.
6. Line a pan with parchment paper.
7. Spread snack mixture evenly into pan; press and pack down tightly.
8. Place mixture in a refrigerator for 2 hours.
9. After 2 hours, remove mixture from pan by lifting parchment paper.
10. Cut into 6 servings.

Names _____    _____

_____    _____

# Healthy Snack Survey

| Survey Question | Definitely no | Sort of | Definitely yes |
|---|---|---|---|
| 1. Is the snack tasty? | | | |
| 2. Does the snack have a good texture? | | | |
| 3. Is the snack easy to carry and to eat "on the go"? | | | |
| 4. Do all the flavors in the snack go well together? | | | |
| 5. Does the snack look good to eat? | | | |
| Count up total votes for each column | | | |
| Multiply number of votes by these numbers | x 1 | x 2 | x 3 |
| **Points scored** | | | |

Add all points together for raw score. _____

6. Is the snack nutritious?
   Yes: add 10 points; No: subtract 10 points _____

7. Does the snack have four or more ingredients?
   Yes: add 3 points; No: add 0 points _____

**Final healthy snack score** _____

# Simple Shelters

## Curriculum Connections

Science: natural resources
Social Studies: Native Americans; people, places, & environments

## Criteria for Product

Shelter will be scored on:

- Standing on its own; sturdier shelters score higher
- Having closed walls or sides, except for the entrance
- Staying in place in the wind; surviving faster wind scores higher
- Staying dry inside during rain

## Constraints for Challenge

- Must use only the materials provided in the group kit, but may barter with other groups for more of the same materials
- Must be small enough to fit on a desktop
- Must complete each stage in the time allowed

### Challenge

Throughout history, people have devised shelters based on weather, geography, and available resources. Students design and build models of simple shelters.

## Materials

### Simple Shelter Materials (for each group)

- 3 large paper grocery bags
- 1 sheet of felt, any color
- 20 craft sticks
- 20 sticks, twigs, or barbecue skewers
- 1 roll of string or twine
- 2 pt. of mud in containers with lids
- 1-gal. plastic bag full of dry leaves
- 1-gal. plastic bag full of dry moss
- 1 shopping bag (to hold materials kit)

### Tools

- Hole punches
- Scissors
- Rulers
- Newspapers (to protect surfaces)

### Additional Materials

- Pictures of traditional Native American shelters
- Chart paper
- Markers
- Pencils
- Plain paper
- Glue sticks
- Digital timer
- 3-speed electric fan
- Watering can and water
- Large, wide, shallow plastic container to catch water (optional)
- Document camera (optional)

# Before You Begin

- Locate and print images of different kinds of Native American dwellings. Good terms for an Internet image search include "wigwams," "longhouses," "teepees," "grass houses," "wattle and daub houses," "chickees," "adobe houses," "igloos," and "brush shelters." Print a different example for each group of three to four students.
- File down any sharp ends on sticks, twigs, and skewers.
- Gather dry leaves and dry moss and put into separate 1-gallon plastic bags.
- Measure mud into pint containers. Artist's clay may be substituted for mud.
- Assemble a materials kit for each group, following the Simple Shelter materials list.
- Make a set of Job Cards (page 13) and a copy of the Test Results (page 53) for each group.
- Make a copy of the Challenge Reflection sheet (page 14) for each student.
- Set up the digital timer where students can check on it.
- Display the Criteria and Constraints where students can see them.
- Choose a safe location to set up the electric fan for wind-testing the shelters.
- Choose an outdoor location for rain-testing the shelters, or use a large plastic container for indoor testing.

Get Set

# 5-Step Process

 **1 ▸ Investigate**
*Session 1*
*45 minutes*

- Ask students to ponder the question, "How does where you live affect the kind of house you live in?" If students have trouble understanding the question, you could ask them how their house might be different if they lived in another part of the country. Give students 1 minute to think silently, and then ask them to turn to a neighbor and take turns sharing their thoughts. Allow 2 minutes for sharing. Call on students to share some of their discussions with the class. Note the main points where everyone can see them, on a board or chart paper, or by using a document camera.

- Remind students that before settlers came to North America from Europe, many different tribes of Native Americans lived here in traditional ways. The kinds of shelters they built depended on the weather, climate, and natural resources of the area, as well as the tribe's way of life.

- Divide the class into groups of three or four students, using whatever grouping strategy is common in your classroom. (Students will form different groups to complete the Challenge.) Provide each group with a large sheet of chart paper, markers, a glue stick, and an image of one type of Native American shelter (each group should have a different type of shelter). Instruct students to glue the picture to the center of the chart paper.

- Tell students that their group has 15 minutes to examine the picture as if they are detectives. They should look for clues about how the shelter was constructed: what materials were used, how the materials are put together, what part of the country it is probably in, how large it is, whether it looks like it took more than one person to build it, how long it probably lasted, how many people could use it at one time, and so forth.

- As students come up with ideas and inferences about the shelter, they should write their thoughts on the chart paper, using markers. Have them think about factors such as climate, weather, geography, and available natural resources. Can they tell something about these factors from looking at the shelters? Can they infer anything about the lifestyle of the tribe—did they traditionally live in one place or move frequently; were they mainly farmers or mainly hunters, or did they do something else; did they live in nuclear or extended families?

- After 15 minutes, invite groups to hold up their papers, one at a time, and share their thoughts. Encourage groups to compare the shelter they observed with the ones observed by the other groups.

**1** *Investigate*
*Session 2*
*20 minutes*

- Briefly discuss the kinds of shelters that the students observed and the ideas that the groups came up with by observing the shelters.

- Display the shelter materials in one kit. Allow students to come up, a few at a time, to view and handle the materials.

- Tell students that their engineering Challenge will be to design their own model shelters using what they've learned by observing Native American shelters. Ask students:

    *Which materials could be used to create walls?*

    *How could you use some of these materials to hold the shelter together?*

    *Is it better to use materials that are stiff or flexible? Why?*

    *Is it necessary to use the same materials for the roof as you use for the walls?*

    *What questions do you have about the Challenge?*

- Review the Criteria and Constraints, as well as the time limits for Brainstorm, Plan, and Build. Explain that bartering means trading one item for another item. Although groups must use only the items in their kits, they may barter with other groups to get more of one kind of item. Explain that students will be in different groups for completing the Challenge, but first, they'll brainstorm ideas about shelters on their own.

**2** *Brainstorm*
*15 minutes*

- As needed, remind students that they will brainstorm independently. Give each student a sheet of plain paper. Instruct students to write Brainstorm at the top. Have them work alone quietly for 10 minutes, thinking of different ways to build a model shelter using the materials in the kit. Students should sketch each idea and add labels showing the suggested material for each part of the model shelter.

- While students brainstorm, walk around the room. Offer starting places for students who are stuck and ask guiding questions to support detailed work. Suggested questions:

    *What materials have properties that are good for this part of the shelter?*

*What size will your model shelter be?*

*What material(s) might work for the walls?*

*How will you connect the shelter materials?*

*Will this shelter stand on its own?*

*What have you done to make sure it can stand up to wind?*

- At the end of 10 minutes, have students review all of their ideas and think about which ones they like best, and why.

## 3 ▸ Plan
### 30 minutes

- Form new groups by considering learning styles; for example, is the student a kinesthetic, visual, or auditory learner, or does she show another strong learning preference? (Do not consider academic level for groups for this Challenge.) Form groups of three to four students so that there are different learning styles in each group, and so that no more than two students from the same Investigate group are in the same Plan group.

- Once students are in groups, tell them that it is time to choose jobs: Speaker, Timekeeper, Materials Manager, and Recorder. Each job suits people with different strengths. For example, students who like to talk usually make great Speakers. Students who are organized make terrific Timekeepers. People who like to work with their hands make excellent Materials Managers. And people who focus on details, keep good notes, and are good artists make fantastic Recorders.

- Explain that each group has students with different strengths. Hand out the Job Card sets. Instruct students to discuss their strengths within the group and then decide who should take each job. (For groups with three students, one student takes two jobs.) Make sure students understand that they will do different jobs in the future, because it's important to know how to do every job, even if they have a favorite.

- After students have chosen their jobs, set the timer for 20 minutes and tell groups to begin planning their shelters. Explain that they may choose one shelter design from Brainstorm that they all agree to, or they may come up with a new design that uses ideas from more than one person for the different parts.

- Remind Timekeepers that it is their job to make sure the group stays on task and that the group completes each stage of the design process on time.

- Empower the Speaker by reminding students that if a disagreement occurs, the Speaker may make the final decision. The Speaker is the only group member allowed to get the teacher involved. The Speaker should also barter with other groups, if it turns out that the group needs more of a particular material. (The trade should not be made at this point, however.)

- Walk around the room, checking in on conversations to make sure that students are being respectful and using good collaborative skills. Ask questions to encourage thinking. Sample questions:

  *How similar or different were your original designs?*

  *How did you decide which materials to use? Why did you decide on those?*

*What properties of the materials did you consider when you decided? Why are those properties important for the shelter?*

*Does everyone agree? Was it difficult to get everyone to agree?*

- After a group has come up with one design, each member draws this version of the group's final plan on a separate sheet of paper. After comparing to be sure all plans look alike, the Recorder draws and labels a large diagram on a sheet of chart paper. This is the version everyone in the group follows while building.

## 4 Build
### 60 minutes

- Decide whether you will allow students to test their shelters for wind and rain during Build. If so, explain some ground rules for using the fan and the watering can. If not, explain that these two criteria will be tested at the end of the Challenge, but not during Build.

- Tell students that it's time to build and test their model shelters. Remind them that everyone helps to build the shelter, as well as do their own jobs. Set the timer for 60 minutes. Remind Timekeepers to keep track of the time and let the group know if they need to speed up to finish.

- Each Materials Manager picks up the materials kit for the group. If the group's Speaker arranged for a trade during Plan, then the Materials Manager takes care of making the trade after picking up the materials kit.

- Remind students to follow the group's plan. Students should work little by little, testing the shelter as they proceed. If something does not work as planned, students may change their shelter. It is the Speaker's job to tell the teacher about the change and explain why it is being made. The Recorder then notes the changes on the plan in a different color.

- As students work, walk around the room, helping to drive each group's progress. Praise students for collaborating well together. Ask guiding questions to support critical thinking skills. Sample questions:

  *Is the shelter coming out the way you thought it would?*

  *Did you need to measure anything? If not, should you? Can you measure without a ruler? How would you do that?*

  *Have you bartered with another group? What did you trade?*

  *Has your group made any changes to your plan?*

  *Have you tested your shelter yet to see if it stands? What did you learn?*

- Store finished shelters in a safe place until it is time for Test & Present.

## 5 Test & Present
### 15 minutes prep, then 5 minutes per group

- Tell the class to think of questions to ask when each group presents. Encourage students to share ideas as a class. List the questions where everyone can see them. Sample questions might include:

  *What materials did you choose for each part of the shelter?*

  *Where is the entrance to the shelter?*

  *What challenges did you face?*

  *How many times did you change your plan?*

  *If you had more time, what would you improve?*

- Guide students to review questions and as a class, choose the best five. Leave only those five questions on the list. Remind students that each group must answer all five questions.
- Call the first group and give the Recorder a Test Results sheet.
- Turn on the document camera or other projection device. Have the Materials Manager place the shelter under the camera so all students can see it projected. Alternatively, have the Materials Manager carry the shelter around so all students have a chance to see it up close.
- While the shelter is being shown, the Speaker answers the five questions chosen by the class.
- When all students have seen the group's shelter, the testing begins. The Recorder completes the Test Results sheet during testing.
  - Students who are in other groups rate the shelter on the first two items on the Test Results sheet. Give the class up to 30 seconds to come to a group consensus on each item for each shelter.
  - To test the shelter against wind, place the shelter in front of the fan and turn the fan on low speed for 10 seconds. If the shelter stays in place and holds together, switch to medium speed for 10 seconds. Repeat the test at high speed.
  - To test the shelter against rain, either go outdoors or place the shelter in a shallow plastic container. Fill the watering can and let it rain down on the shelter for 5 seconds. The Materials Manager then lifts the shelter and reports on the condition of its interior.
- When all groups have presented, give each student a Challenge Reflection sheet to complete.

## Opportunities for Differentiation

**To make it simpler:** Demonstrate for students different ways to connect the materials. For example, mud can be used to hold twigs together, or string can be wrapped around and woven through other materials.

**To make it harder:** Add height and width requirements to the criteria. Find and use figurines to represent people and instruct students to make the shelter large enough to hold a family of four.

Names _____     _____

_____     _____

# Simple Shelter Test Results

| Criterion | Score |
|---|---|

## 1. Stands on its own

Stands easily, score = 3
Stands after a little work, score = 2
Stands, then falls or leans, score = 1
Does not stand on its own, score = 0

_____

## 2. Sides closed, except for entrance

All sides are mostly closed, score = 3
Most sides are closed, score = 2
Some sides are closed, score = 1
All sides are open, score = 0

_____

## 3. Stays in place in the wind

Stays in place in high wind, score = 3
Stays in place in medium wind, score = 2
Stays in place in low wind, score = 1
Moves in any wind, score = 0

_____

## 4. Stays dry inside during rain

Inside is completely dry, score = 2
Inside is damp or is wet in one place, score = 1
Inside is wet in many places, score = 0

_____

## Total Score

_____

# Skyscrapers

## Curriculum Connections

Science: motion & stability
Mathematics: properties of solid prisms

## Criteria for Product

- Must be at least 12 inches tall; taller heights score higher
- Must remain standing and hold together in a model earthquake of 10 shakes; more shakes score higher
- Must hold and support a load of weights placed on top; greater loads score higher

## Constraints for Challenge

- Must not be attached to the cardboard base or anything else
- Bottom must sit within the perimeter of the cardboard base
- Must use only the materials provided
- Must complete each stage in the time allowed

### Challenge

In the late 1800s and early 1900s, more and more people moved into cities to work and live, and civil engineers got creative with the design of housing and office space. Students design and build model skyscrapers.

## Materials

### Skyscraper Materials (for each group)

- 1 pack of 100 index cards (5 × 8 in. preferred; 3 × 5 in. acceptable)
- 1 piece of corrugated cardboard, about 12 × 18 in.
- 1 roll of clear tape
- 1 shopping bag (to hold materials kit)

### Tools

- Scissors
- Rulers
- Yardsticks or metersticks

### Additional Materials

- 100 additional index cards
- Chart paper
- Plain paper
- 3 doz. of the same item to use as weights
- Digital timer
- Digital video camera (optional)
- Method of projecting digital videos (optional)

# Before You Begin

- Collect or purchase packs of index cards, cardboard pieces, rolls of tape, and the items to use as weights. Bolts or heavy washers work well, but any item will do, as long as each one has about the same weight.
- Assemble a materials kit for each group, following the Skyscraper materials list.
- Make a set of Job Cards (page 13) and a copy of Test Results (page 61) for each group.
- Make a copy of the Challenge Reflection sheet (page 14) for each student.
- Write the numbers 1, 2, 3, and 4 on slips of paper so that each group has a set of numbers 1 to 4.
- Set up the digital timer where students can check on it.
- Display the Criteria and Constraints where students can see them.

# 5-Step Process

**1** **Investigate**
*Session 1*
*40 minutes*

- Divide the class into groups of three or four students, using any common classroom grouping strategy. Students will be regrouped for Investigate Session 2 and again to complete the Challenge.
- Give each group four index cards, scissors, and a roll of tape. Tell students they will have some mini-challenges. For the first one, they will have 2 minutes to build the tallest tower they can, using index cards, scissors, and tape. The tower must stand on its own and it may not be taped to the desk. Set the timer.
- When 2 minutes have passed, invite groups to compare results. Discuss as a class what worked and what didn't, when creating a tall tower that stood on its own. Ask students what properties the tallest towers shared.
- Give each group four more index cards and five or six weights. Tell students that this time they will build a structure to support the most weight. Height is not a concern. Follow the same procedure as above: set the timer for 2 minutes, let students work, and when 2 minutes have passed, discuss results as a class. As before, ask students to identify what the strongest structures had in common.
- For the third mini-challenge, tell students they will use four more index cards to build a structure that can withstand an earthquake (modeled by shaking the desk). The structure may not be taped to the desk. Height and holding up weights are not concerns, for this mini-challenge. Set the timer.
- After 2 minutes, have each group take turns shaking their desks to test their structures. Discuss the results as a class, asking students to identify which structures stood the longest during the earthquake, and to think about why those structures were successful.

- Compare the results of all three mini-challenges. Ask:

    *Do the tallest towers have the same properties as the ones that stood the longest during the earthquake?*

    *What about the structures that held the most weights? Do they have the same properties as the other two, or different?*

    *Do you think you could build a structure that can meet all three challenges?*

- Tell students that they will be building structures like that for their next engineering Challenge.

**Investigate**

**Session 2**

*20–30 minutes*

- Place students in different groups from Session 1. Tell students that they will explore how folding index cards into different shapes affects their properties—but they will have more time than they did in the mini-challenges.

- Give each group of students six index cards, tape, and scissors.

- Ask students whether they folded their index cards into any particular shapes during Session 1. Let them describe the shapes using any words they choose, but provide the correct mathematical term for each. For example, if a student says he rolled the index card, agree that it is a roll and introduce the term cylinder. The term for a box is a rectangular solid or rectangular prism, while an index card folded into a triangle is a triangular prism.

- Tell students to work with the index cards to create each of these three shapes—cylinder, rectangular prism, and triangular prism. Students should build both long and short versions of each of the three shapes. Assist any groups that have difficulty.

- When the students have built both long and short versions of each, have them test the shapes for weight and earthquakes, as they did in Session 1.

- Tell students that civil engineers also use flat, two-dimensional materials to build three-dimensional structures, such as skyscrapers. Often, the shape is repeated again and again throughout the structure. They test the materials and shapes that are used to build the structures to be sure that the buildings will stand up under heavy loads and during earthquakes, and so tall buildings can stand on their own.

- Tell students that they will put all their knowledge to work as they design skyscrapers, using the same materials and tools they've been investigating—index cards, tape, and scissors.

- Review with students the Criteria and Constraints, as well as the time limits for Brainstorm, Plan, and Build.

**2 ▶ Brainstorm**

*15 minutes*

- Provide each student with a sheet of plain paper. Instruct them to fold the paper in half vertically and again horizontally, creating four sections when unfolded. Have them write Brainstorm at the top.

- Explain to students that for this Challenge, each skyscraper's score will depend on how it compares to the other skyscrapers on each criterion. If a minimum criterion is not met, then the score for that criterion is zero. But skyscrapers that exceed criteria will score additional points. Show the Test Results sheet as needed to explain the scoring.

- Tell students they will now brainstorm four different options for building a skyscraper that meets all the Criteria, using only the available materials.

- Set the timer for 10 minutes and instruct students to begin drawing.

- Ask students:

  *What shapes are you planning to use? Why did you choose those shapes?*

  *Will you use short shapes, long shapes, or both?*

  *Will the skyscraper be the same width as it gets taller? Why might you want it to have different widths at different heights?*

  *How will you hold the pieces together?*

- After 10 minutes, instruct students to circle one favorite idea.

**3 ▶ Plan**

*30 minutes*

- Organize groups of three to four students who work well together. Try to place friends in the same groups for this Challenge.

- Provide each group with a set of Job Cards and a set of number slips 1 to 4. Students in each group take turns choosing a number. The person who chooses 1 gets to pick the job she would like to do. Job choice continues in order until the person with number 4 takes the remaining job. Students may not trade jobs, but groups of three may decide who gets the leftover job. Remind students of their job responsibilities (page 12).

- Tell students that they will work together to create a group plan that everyone will help to build. They should take turns, sharing their favorite ideas and piggy-backing on someone's idea that is similar to what they have planned.

- Let students begin planning as soon as they've chosen jobs. When all groups are ready, set the timer for 20 minutes. As students talk, stop by and check with each group. Ask guiding questions. Sample questions:

  *Which jobs will each of you take on?*

  *How will you build the skyscraper? Will you make all the cards into shapes first, and then build, or will you break up the jobs into shape makers and skyscraper builders?*

  *How will your skyscraper hold the weights?*

  *How will you test strength as you build without using the actual weights?*

## 4 ▸ Build

*60 minutes*

- Have the Materials Managers pick up a kit for each group. Give each Recorder a Test Results sheet so the group has the criteria for reference during Build.

- Remind students that they need to build on the cardboard pieces, but they may not attach the skyscrapers to the cardboard. (The cardboard pieces make it easier to move the skyscrapers for storage and testing.) Instruct Timekeepers to call out the remaining time periodically and to make sure that everyone is contributing.

- When all groups are ready, set the timer for 60 minutes. As students work, check in with groups, ask guiding questions, and allow students to explain their reasoning. Suggested questions:

  *How did you decide on your plan? What was your procedure?*

  *What job does each person have?*

  *How is each person helping to build the skyscraper?*

  *Why did you decide to build with those shapes?*

  *Is there a place on top for the weights to be added?*

  *Which part will help the skyscraper stand up during an earthquake?*

- Store finished skyscrapers in a safe place until it is time for Test & Present.

## 5 ▸ Test & Present

*15 minutes of discussion, then 5 minutes per group*

- Display the questions in the box "Skyscraper Presentation Questions" where the whole class can see them. Give each group time to discuss their answers.

- Tell students that each group will conduct three tests on their skyscraper. First, they use a yardstick to measure the height (in inches) of the skyscraper. Next, they shake the desk back and forth up to 20 times, counting the shakes as they go and stopping when the skyscraper starts to fall. Last, they place weights on top of the skyscraper, one at a time, until it begins to buckle.

- Call on groups, one at a time, to present. If you plan to have students make videos, hand the Recorder a camera. Explain that the Recorder should record only the testing parts of the presentations, for no more than 3 minutes of video time. Make sure the Recorder also has a copy of the Test Results sheet, distributed earlier.

- Whether the presentations are recorded or presented without recording, the procedure is the same.

  - The **Speaker** answers the Presentation Questions and does most of the speaking.

  - The **Materials Manager** does the testing.

  - The **Recorder** writes down the results (and makes the video, if desired).

  - The **Timekeeper** assists the Materials Manager and makes sure the total presentation is no longer than 5 minutes.

- When all groups have finished, line up the skyscrapers by size to determine the points to be awarded to each for item 1 on the Test Results sheet. Then compare group results for earthquakes and weight, and score items 2 and 3.
- Have the Recorders add up the scores for their skyscrapers and determine which skyscraper has the lowest score.
- Give each student a Challenge Reflection sheet to complete.
- If allowed by your district's policies, and if you have permission from parents, publish the videos to an approved, secure website. Show students how to find and play the videos.

## *Opportunities for Differentiation*

**To make it simpler:** Demonstrate for students how to create strong shapes from index cards. Allow students to tape skyscrapers to a piece of cardboard as a base.

**To make it harder:** Limit each group to 1 yard of tape or completely remove tape from the kit. Increase the weight of the minimum load. Increase the minimum height and do not test the other criteria if the skyscraper does not meet height requirements.

# Skyscraper Presentation Questions

Read each question. Then say the answer for your group.

1. What was your process for building the skyscraper?

2. Who did each task in the process?

3. How did you create the shapes?

4. How tall is your skyscraper?

5. How many times were you able to shake the desk (table) before the skyscraper fell?

6. How many weights did your skyscraper hold? Did any part of your skyscraper buckle under the weight? Which part?

7. What part of your skyscraper would you improve? How?

Names _____    _____

_____    _____

## Skyscraper Test Results

| Height (inches) | Quakes Withstood | Weights Held |
|---|---|---|
|  |  |  |

First complete all tests, and then figure out the scoring. For each criterion, compare all skyscrapers. Award points based on rank: 1 point for first place, 2 points for second place, and so forth. In case of a tie, award both skyscrapers the same number of points.

# Criterion                                                    Score

### 1. Height                                                    _____
Rank from tallest (score = 1 point) to shortest.
If it is less than 12 inches tall, add 5 points to the rank score.

### 2. Ability to withstand earthquakes                          _____
Rank in order from the greatest number of shakes to least.
If it falls before 10 shakes, add 5 points to the rank score.

### 3. Ability to support a load                                 _____
Rank in order from the most weights held (score = 1 point) to least.
If it cannot hold any weights, add 5 points to the rank score.

### Total Score (lowest is best in class)                        _____

# Model Cars

## Curriculum Connections

Science: motion, speed, variables
Mathematics: measuring distance & time, comparing numbers, division, rates

## Criteria for Product

- Speed—faster cars score higher
- Distance—cars that travel farther score higher
- Appearance—better-looking cars score higher

## Constraints for Challenge

- Must use only the materials provided
- Must complete each stage in time allowed
- Model car must be tested three times on the course provided

### Challenge

Model cars are fun! They are also great vehicles for learning about the effects of design on product performance. Students work as mechanical engineers to design, build, and test model cars.

## Materials

### Model Car Materials (for the class)

- 50 manila folders
- 100 pipe cleaners
- 100 barbecue skewers
- 100 straws
- 100 coffee stirrers
- 100 toothpicks
- 100 craft sticks
- 8 egg cartons
- 25 small paper cups, 6-oz. size
- Assorted cardboard boxes
- Paper towel and bath tissue tubes
- 1 bag of hard Lifesavers candy rolls
- 1 bag of soft Lifesavers candy rolls
- 100 recycled CDs
- 1 box of large paper clips
- 6 rolls of masking tape

### Tools

- Rulers
- Scissors
- Tape measures

### Additional Materials

- Computer with Internet access (optional)
- Projector for computer (optional)
- Chart paper
- Markers
- Digital timer
- Ramp and hill setup (see "Before You Begin")
- Toy cars of different sizes

# Before You Begin

- Purchase or locate items to use to build a ramp and hill as shown on the Test Results sheet (page 67). Blocks, books, boards, and pieces of cardboard are good options. The ramp should be 1 foot wide, 9 inches tall, and 2 feet long. The hill is 1 foot wide, 5 inches tall, 1 foot up the hill, 4 inches across the top, and 1 foot down the hill. The ramp and hill are a foot apart. All measurements are approximate.

- Practice setting up the ramp and the hill, test them using toy cars, and tweak the setup as needed. Decide whether you'd like to improvise additional ramps for use during Investigate Session 2.

- Go to the BBC website (www.bbc.co.uk), search "forces and movement," and then click on the "Schools Science Clips Online lesson plan." Once there, click on the link for "Activity." Bookmark the car simulation for use as a class activity or for student use at computer stations.

Get Set

- Make a set of Job Cards (page 13) and a copy of the Test Results (page 67) for each group.

- Make a copy of the Challenge Reflection sheet (page 14) for each student.

- Set up the digital timer where students can check on it.

- Display the Criteria and Constraints where students can see them.

# 5-Step Process

**1** *Investigate*
*Session 1*
*45 minutes*

- Ask students to think about the following question, or a rephrasing of it: "What can affect the speed and distance of a rolling vehicle?" Do not discuss the question yet.

- Review or define the term *variable*. Explain that the things that affect how a vehicle rolls are examples of variables. Draw a graphic organizer (Venn diagram, T-chart, or another type) where the class can see it, for the purpose of comparing and contrasting the effects of some variables on speed and distance.

- Invite students to share ideas of variables that might affect either the speed of the car, the distance it travels, or both (for example, the car's weight, how easily its wheels turn, how hard it's pushed). Write these variables on the organizer in their correct places.

- Project the simulation you bookmarked earlier and use it as a class activity, or arrange for students to use it at computer stations. In this interactive online activity, students investigate how the speed and distance that a car rolls are affected by variables, including the size (weight) of the car, the force pushing it, and the starting height of a ramp.

- Allow students to explore the simulation, following the directions. Check in as they work, asking questions to be sure they're thinking about the effects of each variable on the car's speed and distance.

- When students have finished exploring, bring up the graphic organizer again. As a class, compare what students have learned to their original thoughts. Make adjustments to the organizer or start a new one to reflect new knowledge and understanding.

**1** **Investigate**
*Session 2*
**30 minutes**

- Set up the test ramp but not the hill. Gather students around the ramp. Tell students they will now do some real-life tests of variables using toy cars, a ramp, and weights. You may choose to provide additional simple ramps and allow students to work in small groups of their own choosing.

- Ask students to think of different ways to test the toy cars. They can change one variable each trial. For example, the first test might be a small car going down the ramp with no push. The second test would change one thing, such as adding a gentle push, placing a weight on the car, raising the ramp, or using a larger or smaller car.

- Allow about 15 minutes for testing the cars. Question students as they investigate—do their results match the simulation from Session 1?

- Next, add the hill to the end of the ramp and ask students how they could get a vehicle to go down the ramp and up and over the hill. Do not allow them to test this time. They should just think about it. They'll have a chance to test their ideas with their own models.

- Tell the students that rather than continuing to work with the toy cars, they will design and build model cars of their own. Their goal is to build a car that travels as far as possible in the shortest time possible, and that looks good. Read the Criteria and Constraints aloud, and post them where students can see them. Review the Test Results sheet so students understand how their cars will be scored.

- Display the model car materials on a table. Invite small groups of students to take turns coming up to view them. Ask students:

  *What parts does a model car need?*

  *What can you use to hold the wheels on the car?*

  *How can you make sure that the wheels will turn?*

  *Do you have any questions about this Challenge?*

- Review with students the Criteria and Constraints, as well as the time limits for Brainstorm, Plan, and Build.

**2** **Brainstorm**
**30 minutes**

- Make groups of three to four students with different learning styles (kinesthetic, visual, auditory, or others). Give each group a set of Job Cards.

- Explain that each group needs to decide who does which job. Guide students to think about the things each group member is good at. Support students as they decide which jobs each member should do. In groups of three, one student needs to take two jobs. Remind students of the responsibilities of each group job (page 12).

- Tell students that the Materials Manager will make final decisions about which materials to use if the group can't decide. As always, the Timekeeper is responsible for keeping everyone on task so the group finishes on time.

- Tell students that for this Challenge, they will brainstorm as a group. Remind students that Brainstorm is not a time to judge or to choose the exact materials. It is a time to think of all the possibilities and get them down on paper.

- Give each group a sheet of chart paper and markers. Instruct the groups to first list the different parts of a model car and then think of as many options as they can for making each part. They should list each part and then write all the materials that could be used to create that part. Set the timer for 10 minutes and tell students to begin.

- Walk around the room, listening in to conversations and checking to be sure students are brainstorming, not evaluating ideas or making decisions. Encourage all students to offer ideas so that all voices can be heard.

- When the timer goes off, tell students it is time to stop the brainstorming and to put down the markers.

## Plan
### 30 minutes

- Tell students that during Plan they will evaluate the ideas from Brainstorm and make decisions about the best way to use the materials to build a car that meets the criteria.

- The Recorder jots down ideas as the group plans. Once a decision is made, the Recorder asks for a fresh sheet of chart paper to draw the plan. The final plan should use a different color for each building material. The rest of the group helps with the final plan by writing labels and adding a key to the color codes.

- Set the timer for 20 minutes. While students are working, walk around the room and talk with each group. Ask guiding questions. Sample questions:

    *What items will you use for wheels? How are you attaching them?*

    *What are you using for the axles? Why did you choose that?*

    *How can you get the car to go faster?*

    *When you build, what part will each of you work on?*

    *Are you stuck on anything? What can I help with?*

## Build
### 60 minutes

- Tell students it's time to work as a group to build their model cars. Set the timer for 60 minutes. Have Materials Managers collect the items listed in their groups' plans.

- Make sure students remember that if there are any changes to their plan, the Speaker needs to notify the teacher and get permission before continuing, and the Recorder needs to note the changes on the plan. The Timekeeper needs to make sure that the group still finishes on time, even with the change.

- As students work, continue to offer support. Suggested questions:

    *Is every person doing a fair share of the building?*

    *Are all the materials working out the way you imagined?*

    *What are you doing right now? How is that helping?*

    *Are you working on one piece at a time or are you building parts separately and bringing them together at the end?*

    *Have you tested your car yet? Did you make changes after testing it?*

- Store model cars in a safe place until Test & Present.

## 5 ▸ Test & Present

*10 minutes of discussion, then 5 minutes per group*

- Ask students to list some questions that would be good to ask of each group. Write all questions where students can see them.
- As a class, review all questions and choose the best five. Leave only those questions on the list; these are the questions that each group must answer during their presentations. Sample questions:

    *How did you come up with the idea for this design?*

    *What properties does that material have? Is that why you chose it?*

    *What is your favorite part of your car?*

    *Did you have any surprises while you were building?*

- Set up the ramp and hill. Invite one group at a time to test their car, following this procedure.

    - The **Speaker** answers the five questions about the group's work.
    - The **Timekeeper** works the stopwatch to measure the time from when the car starts until it stops.
    - The **Materials Manager** places the car at the top of the ramp and either lets it go or gives it a push.
    - The **Recorder** completes the Test Results sheet, with help from the rest of the group.

- Each group tests their car three times and records the results.
- When all cars have been tested and the average speed calculated, rank the cars in order. For example, the car that travels the greatest average distance scores 1 point; the second greatest average distance scores 2 points; and so forth. (Note: If your students have not yet studied division, have them score "Least Average Time" rather than "Greatest Average Speed.")
- As a class, discuss and rank the cars according to appearance. The best-looking car has a score of 1, next best has a score of 2, and so forth.
- Have students add up total scores for their cars. Because the scores are based on rank, the lowest-ranking car is the best in the class.
- Distribute a copy of the Challenge Reflection sheet to each student after all presentations are finished.

## Opportunities for Differentiation

**To make it simpler:** Remove the hill from the test setup, raise the end of the starting ramp, or both. Provide all groups the same basic car design: wheels and axles attached to a body. Then, allow students to modify the already-built car.

**To make it harder:** Require the cars to start rolling without a push. Or challenge students to get the car to stop at the top of the hill, rather than roll over it. Or add a requirement that the cars carry cargo (weights) and run additional tests to determine how much cargo can be carried securely.

Names _____    _____

_____    _____

# Model Car Test Results

| Test | Distance (inches) | Time (seconds) | Speed (inches per second) |
|------|-------------------|----------------|---------------------------|
| 1 | | | |
| 2 | | | |
| 3 | | | |
| Average | | | |

1.  Draw the car on the diagram to show where it stopped in each test.

Ramp                                    Hill

2.  For each criterion, rank all the cars in the class from first to last. Your car's rank is your score for that criterion. Ties score the same.

3.  Add up all the scores. Lowest total score is best model car overall.

| Criterion | Our Car's Score |
|-----------|-----------------|
| Greatest Average Distance | |
| Greatest Average Speed | |
| Appearance (class agreement) | |
| **TOTAL** | |

# Board Games

## Curriculum Connections

Science: define a problem, design a solution
Mathematics: classifying; two- and three-dimensional shapes

## Criteria for Product

Game will be scored by classmates on each of the following features:
- Catchy name—something attention-getting
- Fun factor—must be fun to play
- Clear instructions on how to play—written, drawn, or both
- Challenge level—not too easy, not too hard
- Reward—something that players receive as part of play
- Variation—the game is different each time it is played
- 3-D feature—at least one part or piece is not flat and is important to the game

## Constraints for Challenge

- Must use only the materials provided
- Must stay within a budget of $5.00 for manufacturing costs
- Must require at least two players
- Must be sturdy enough to hold together through five playtests
- Must complete each stage in the time allowed

### Challenge

Games have challenged and entertained us for millennia—and new games are invented every year. Students design, build, and playtest prototypes for new board games.

## Materials

### Board Game Materials (for the class)

- Construction paper
- Plain paper
- Modeling clay
- Small clear take-out containers with lids
- Cardboard boxes
- Cardboard tubes
- Pipe cleaners
- Assorted small jars
- Cans
- Paper bags
- Aluminum foil
- Marbles
- Dice
- Paper clips

### Tools

- Scissors
- Hole punches
- Glue gun (teacher use only)
- Crayons
- Permanent markers
- White glue
- Glue sticks

### Additional Materials

- Variety of common board games
- Pretend money, $5.00 for each group
- Envelopes
- Box for the "cash register"
- Chart paper
- Markers
- Plain paper
- Pencils
- Sticky notes
- Digital timer

Note: *Provide materials in quantities that are readily available.*

# Before You Begin

- Request donations of clean recyclables from families and shop for other materials as needed.
- Gather examples of popular board games that your students may have played. Try to include a variety with obvious differences, for example, Scrabble, Sorry, Hi-Ho Cherry-O, Operation, Clue, Checkers, Chess, Go, and so forth.
- Divide modeling clay into sticks or roll into balls about 1½ inches in diameter.
- Make signs for the price of each material or copy the list on page 74, making any necessary adjustments for available materials.
- Set up a box or other container to serve as the "cash register" for the materials shop.
- For each group, set up an envelope with $5.00 of pretend money.
- Make a set of Job Cards (page 13), a copy of the Materials Price List (page 74), and a copy of the Test Results (page 76) for each group.
- Make a copy of the Challenge Reflection sheet (page 14) for each student.
- Make several copies of the Board Game Survey (page 75), as each student will need as many copies of the survey as there are groups in the class.
- Set up the digital timer where students can check on it.
- Display the Criteria and Constraints where students can see them.

# 5-Step Process

**1** *Investigate*
*Session 1*
*20 minutes*

- Provide each student with five sticky notes. Have a whiteboard or sheet of chart paper or poster board available that the sticky notes will adhere to.
- Ask students to recall board games they've played and list up to five favorite games, one on each sticky note. (You may need to reinforce that the question is about board games, not video games.)
- Call on students to name the games written on their sticky notes, and then to come up and place their sticky notes on the board. Require each student to share the name of a game that is not already on the board. After all students have had a chance to share, ask students to add sticky notes for any games that are not yet on the board.
- Tell students that next, they're going to sort the games into categories based on what they are about, how they are played, or something else they have in common. Encourage students to think of ways to re-sort the games into different categories. The goal is to show students that there are many ways to look at the features of the games. Sorting suggestions:

  *Does the game include cards?*

  *Does a player have one game piece or multiple game pieces?*

  *Do the game pieces move on the board or do they stay in one place?*

  *Is the game board two-dimensional or three-dimensional?*

*Investigate*
*Session 2*
*30 minutes*

- Bring out some actual board games for students to examine. You may choose to have students play the games or simply take them out, look at them, and discuss their properties and features in small groups or as a class.
- Whether students are playing or discussing the games, explain that they are not just considering how fun they are. Instead, they will analyze the games the way a game designer would. As students investigate, ask:

  *What are some ways that different games allow a player to make a move?*

  *What are the rewards and punishments within the game? In other words, what happens during the game that players enjoy and what happens that they don't enjoy?*

  *Which features of this game do you enjoy? Which do you not enjoy? Why?*

- Tell the students they've been thinking like game designers. Their next Challenge is to design, build, and playtest new board games.
- Review with students the Criteria and Constraints, as well as the time limits for Brainstorm, Plan, and Build.

**2** *Brainstorm*
*30 minutes*

- Display the board game materials on a table.
- Talk to students about their experiences with buying things. Tell students that each material for this Challenge has a price. Each group will be given $5.00 to buy materials, which means the group's game prototype can cost no more than their budget of $5.00. Give each group a copy of the Materials Price List.
- Invite students to come to the table in small groups to look at the materials and their prices. Ask students:

  *How could you use these materials to build a board game?*

  *Can you change the shape of any of the materials?*

  *Can you place two or more materials together to create something new?*

  *Which materials would be good for making the board?*

  *Which materials would be good for making a three-dimensional feature?*

- Give each student a sheet of plain paper and instruct her to fold it in half vertically and then again horizontally, creating four sections when unfolded. Have her write Brainstorm at the top.
- Instruct students to work quietly by themselves for 10 minutes. Their task is to think of a design for a board game and draw it on paper.
  - In the upper left-hand box, students draw the game board.
  - In the upper right-hand box, students draw the playing pieces.
  - In the bottom left, they sketch the cards, dice, spinner, popper, or other device that determines how players move. (If their game does not have this, then they may leave it blank.)
  - In the bottom right, they write directions for playing the game.

- Encourage students to add labels that show the materials used.
- While students brainstorm, walk around the room. Offer starting places for students who are stuck and ask guiding questions to support detailed work. Suggested questions:

  *Will your three-dimensional feature be a hill or valley on the board, a dice-rolling device, or something else?*

  *Which material will you use to make that part of your game? Why is that a good choice?*

  *How will you decorate your game board? How will the decorations help the players?*

  *How will players know how many spaces to move or where to place their pieces?*

  *How big will your game be?*

- If time permits, students may come up with more than one design. If they do, then they need to decide which one is their favorite.

### Plan

**30 minutes**

- Ask students to stand in the appropriate groups as you call out descriptions of board games. You may use the following descriptions or modify them to better suit your class's designs from Brainstorm:
  - Game follows a simple path that moves in one direction
  - Game has multiple paths
  - Game has no path
- Once students are in large groups, move from group to group, creating groups of three to four students based on similarities in design, such as using cards, dice, or poppers.
- Explain that each group has students with game designs that have something in common with each other. Hand out Job Card sets and instruct students to take 2 minutes to decide which student should take each job. Ask students to review, out loud, the responsibilities of each group job (page 12).
- Once students have chosen jobs, set the timer for 20 minutes and let groups begin planning.
- Tell students to place their Brainstorm pages where everyone in the group can see them. Ask them to look for ways in which their designs are similar. They should start their group plan by designing the features that their four games have in common.
- Walk around the room, checking in on conversations to make sure that students are being respectful, staying on task, doing their chosen jobs, and working well together. Sample questions:

  *How similar or different were your Brainstorm designs?*

  *How did you decide which materials to use? Why did you decide on those?*

  *Is everyone in agreement? Was it difficult to get everyone to agree?*

  *What material did you choose for the cards? For the game pieces?*

*Will those materials hold up while people play the game multiple times?*

*Have you considered all the Criteria and Constraints?*

*How will the cards stay stacked?*

- After a group agrees to a plan, instruct students to write up a shopping list, look at the prices for the materials, and add up the total for making one copy of the game. If their design is over budget, then they need to change their design.

- Once the group has a final plan, the Recorder divides a sheet of chart paper into four sections, then assigns one member of the group to draw each of the categories that were sketched during Brainstorm: game board, playing pieces, device that determines how players move, and instructions for play.

- After the group has finished drawing up their plan, all group members must review and sign off on all four parts of the plan. If they do not agree on part of the plan, then they need to discuss that part again and make changes to the plan before signing off.

## 4 Build

### 60 minutes

- Tell students it is time to build their games. Set the timer for 60 minutes. Materials Managers bring their group's shopping list to the table and receive a money envelope to purchase materials. They give their money to purchase materials to the teacher or another adult who is serving as shopkeeper.

- Remind groups to follow their plans. If a plan changes, the Materials Manager must figure out if the budget will cover the change. Then the Speaker must get permission from the teacher, and the Recorder must draw the change on the plan in a different color. This process is time-consuming, so the Timekeeper needs to let the group know right away if there is not enough time to even ask for a change. This process will help reinforce that it is better to make changes during the Plan stage.

- As students work, walk around the room, checking the progress of each group. Praise students for working together well. Sample questions:

    *How is each person participating? How did you decide who would make each part of the game?*

    *Why did each person take each group job?*

    *Is the game coming out the way you planned it?*

    *Have you made any changes to your plan?*

    *Who is writing the instructions?*

- Store finished games in a safe place until it's time for Test & Present.

## 5 Test & Present

*5 minutes per group to present, then 50–70 minutes for playtest*

- Ask the first group to come up to the front of the room with their game. Ask the Speaker to explain how the game is played. Then the Materials Manager demonstrates how the game meets each of the Criteria for Product.

- Invite students in other groups to ask questions, which the Speaker answers. Allow up to five questions.

- When all groups have finished presenting, have each group set up its game for play. Explain that groups will playtest the games created by the other groups. Remind students to be respectful of their classmates' creations and to handle the games gently, because these are the only prototypes of these games available, anywhere.

- To playtest the games, allow each group 5 to 10 minutes of playtime. Then, each student takes up to 2 minutes to complete a Board Game Survey about the game they just played. The Timekeepers make sure everyone in their group finishes each survey on time.

- When all surveys are complete for the first test, collect them for each game and store them until the end of playtesting. Students then rotate to another game for the next test. (For a 21st-century approach, students could create and complete surveys using an online survey creation tool.)

- Once all games have been played and all surveys completed, the Recorder for each group receives the surveys for his group's game, as well as a Test Results sheet. Recorders should work with the others in their group to figure out the average score for each criterion and complete the Test Results sheet. If students do not yet know how to divide, have them total the scores without finding the average.

- Provide each student with a Challenge Reflection sheet to complete.

## Opportunities for Differentiation

**To make it simpler:** First, provide templates for basic game boards and allow students to choose from those. Second, ask students to choose dice, a spinner, or cards in order to move the play forward. Last, allow students to choose items from around the classroom to use as playing pieces, such as small blocks, erasers, and so forth.

**To make it harder:** Include more specific requirements in the Criteria, such as "must include dice and cards," "must require players to do something unexpected during play," or "must include the chance of a random event that can set back whoever is ahead."

# Board Game Materials Price List

Construction paper ............................................... 25¢ each

Plain paper .......................................................... 10¢ each

Modeling clay, sticks or 1½-in. diameter balls ....... 15¢ each

Plastic take-out containers ...................................... 15¢ each

Cardboard boxes ................................................... 10¢ each

Paper towel and bath tissue tubes .......................... 5¢ each

Pipe cleaners ........................................................ 3¢ each

Jars ...................................................................... 25¢ each

Cans ..................................................................... 10¢ each

Paper bags ............................................................ 5¢ each

Aluminum foil ....................................................... 3¢ per foot

Marbles ................................................................ 2¢ each

Dice ..................................................................... 5¢ each

Paper clips ............................................................ 1¢ each

Name of Game _____

Name of Reviewer _____

(Names are needed in case there are questions about the survey responses.)

# Board Game Survey

Circle your choice for each item after playing the game.

**1. How catchy is the name?**

| 1 | 2 | 3 |
|---|---|---|
| Not very catchy | Catchy enough | Very catchy |

**2. How fun is this game to play?**

| 1 | 2 | 3 |
|---|---|---|
| A little fun | Fun | Very fun |

**3. How clear are the instructions?**

| 1 | 2 | 3 |
|---|---|---|
| A little confusing | Okay | Very clear |

**4. How is the challenge level of this game?**

| 1 | 2 | 3 |
|---|---|---|
| Way too easy or too hard | A little too easy or too hard | Just right |

**5. How do you rate the reward(s) in this game?**

| 1 | 2 | 3 |
|---|---|---|
| Okay | Worth getting | Awesome |

**6. Will the game be different the next time it's played?**

| 1 | 2 | 3 |
|---|---|---|
| It will be the same. | It will change a little. | It could be very different. |

**7. Rate the importance of the 3-D feature(s) in playing this game.**

| 1 | 2 | 3 |
|---|---|---|
| Could still play without it | It helps | Can't play without it |

Note: If the game has no 3-D feature, score zero for item 7.

Names _____   _____

_____   _____

# Board Game Test Results

Name of Board Game _____

Use the completed Board Game surveys to find your game's scores.

For each criterion, find your game's average score. For each numbered item on the survey sheet, do the following:

- Add together the scores from all the surveys for that numbered item.
- Divide the sum by the number of survey answers for that item.
- Write the average score for each criterion in the appropriate box in the table.

| Criterion | Average Score |
|---|---|
| 1. Catchy name | |
| 2. Fun factor | |
| 3. Clear instructions | |
| 4. Challenge level | |
| 5. Reward | |
| 6. Variation | |
| 7. 3-D feature | |
| **TOTAL SCORE** | |

# Ski Lifts

## Curriculum Connections

Science: simple machines, forces
Mathematics: average (mean)

## Criteria for Product

- Must have a chair or gondola to carry passengers
- Must safely carry a load of at least two passengers (weights) from the floor to a desktop; carrying more passengers scores higher
- Must complete the trip from the floor to the desktop in 2 minutes or less, starting from 4 feet away

## Constraints for Challenge

- Must use only the materials provided
- Must be human-powered
- Must choose only two items from each set of materials
- Classroom furniture may not be part of the lift, but the ends of the lift may be attached to furniture for stability
- The lift's chair or gondola must not be touched once the load is placed
- The lift does not have to return its chair or gondola to the floor

**Challenge**

With winter comes snow in many parts of the country, and snow means snow sports. Students design and build model ski lifts to carry passengers up a mountain.

## Ski Lift Materials (for the class)

**Set 1**
- Egg cartons
- Small cans
- Cereal boxes
- Assorted shapes and sizes of plastic bottles
- Lunch-size paper bags
- Paper cups, any size
- Paper plates, any size

**Set 2**
- Pipe cleaners
- Rubber bands
- Yarn
- String

- Narrow ribbon
- Fishing line

**Set 3**
- Newspaper
- Waxed paper
- Plastic wrap
- Aluminum foil

**Set 4**
- Craft sticks
- Plastic spoons
- Cardboard tubes
- Straws
- Paper clips
- Twist ties

**Tools**
- 3 rolls of masking tape or duct tape
- 2–4 bottles of glue
- Hole punches
- Scissors
- Rulers
- Staplers

**Additional Materials**
- Toy lifting machines
- Construction paper
- Chart paper
- Markers

- 6–10 bolts or washers to use as "passengers"
- Pencils
- Plain paper
- Stopwatch
- Tape measure
- Digital timer
- Computer with Internet access (optional)
- Projector for computer (optional)

# Before You Begin

- Contact families to arrange for donations of clean recyclables and demonstration items.
- For demonstration, obtain a variety of toys that lift loads, such as toy cranes, dump trucks, backhoes, and parking garages.
- Search online for videos that show examples of different ski lifts in action, including the main parts of the machine. Good search terms include combinations of the words "ski lift," "chair lift," "rope tow," "mountain," "gondola."
- Make labels (Set 1, Set 2, Set 3, Set 4) for displaying the sets of Ski Lift materials.
- Make a set of Job Cards (page 13) and a copy of Test Results (page 82) for each group.
- Make a copy of the Challenge Reflection sheet (page 14) for each student.
- Set up the digital timer where students can check on it.

# 5-Step Process

**1 Investigate**
**Session 1**
**45 minutes**

- Write the words *simple machine* where they can be seen and discussed as a class. Invite students to share words that come to mind when they think about the word *machine*. Write their words randomly all around the word machine, at angles, in different colors, and in different kinds of letters. Explain to students that this is a word splash, and it's a great way to draw on what they already know about a topic.
- Put students in pairs and allow 2 minutes for partners to discuss any connections they can make between any of the words in the word splash. Invite pairs to share ideas from their discussions.
- Display the collection of toy machines and allow students to explore them. Then ask if they can associate any of the word-splash words with these objects.
- Review with students the idea that machines help to make work easier, and explain that there are six basic simple machines: lever, wheel and axle, pulley, wedge, inclined plane, and screw. Ask students to share common household examples of these machines.

**1 Investigate**
**Session 2**
**30 minutes**

- Tell students that a ski lift is one example of a real-world engineering solution that uses a combination of simple machines. Ask students if they have ever seen or experienced a ski lift in winter or in summer. Allow students to share their experiences.
- Show students the videos and images you located that demonstrate different kinds of ski lifts as well as the main parts of a ski lift (chair or gondola, cable, towers, power source). Explain that some ski lifts are open chairs that carry just a few people. Other ski lifts are enclosed gondolas or trams, some of which can carry whole groups.

- Tell students that they will use what they know about simple machines to design a model ski lift. Show students the Test Results sheet so they know how their ski lifts will be judged.
- Review with students the Criteria and Constraints, as well as the time limits for Brainstorm, Plan, and Build.

## ❷ Brainstorm
**30 minutes**

- Before beginning Brainstorm, place the sets of materials on a table where students can see them, with the signs Set 1, Set 2, Set 3, and Set 4. Place the tools and "passengers" (weights) out as well.
- Invite students a few at a time to come look at the materials. Point out the different sets, and explain that they will be allowed to use only two kinds of materials from each set in their ski-lift design. Ask students:

  *What parts of the ski lift do you need to include in your design?*

  *What simple machines might be helpful as part of a ski lift?*

  *Can you change the shape of any of these materials?*

  *Can you use one material in more than one way?*

  *Which set of materials would be best for designing a chair?*

  *Which set of items might be best for pulling or pushing the chair or gondola, or for reducing friction?*

- After students have viewed the materials, give them each 2 sheets of plain paper. Tell them to fold both sheets in half vertically and then again horizontally, creating four boxes, and to label one sheet Brainstorm. Have them label the other sheet Materials List and label its four boxes Set 1, Set 2, Set 3, and Set 4.
- Have students work for 10 minutes to design and draw four different ski lifts. Each ski-lift design needs to include two materials from each set, for a total of eight materials.
- During Brainstorm, walk around the room and check in with each student. Ask guiding questions to help students who are stuck, and to encourage students who are rushing through to slow down and think about details. Suggested questions:

  *What gave you the idea to do it that way?*

  *Which material will you use for that? Is it the best choice? Why?*

  *What experience do you have with these materials that helped you choose them?*

- After 10 minutes, have each student circle one favorite design and its materials.

## ❸ Plan
**30 minutes**

- Ask each student to look at her favorite ski-lift design from Brainstorm. How does it work? Does it use a simple machine? If so, which one (pulley, inclined plane, or another simple machine)? Ask the class to divide into two groups—pulleys and not pulleys— based on the simple machines used by their ski-lift design.

- Invite students to further divide themselves into groups of four (or three, if four is not possible), based on additional similarities. For example, the "pulley group" could subdivide based on the materials used for the pulley in each design. The "not-pulley group" could subdivide based on the simple machine used in each design. As needed, subgroups can be further divided by similar materials. Assist any students who aren't sure where they belong.

- Pass out Job Cards to each group and review the responsibilities of each job (page 12). Let groups choose jobs in any way they like.

- Tell students to place their Brainstorm sketches on the table so that everyone in their group can see all the designs. Ask them to look at the designs for similarities and start there to create a group plan. Set the timer for 20 minutes of Plan time.

- Walk around the room. Check in with groups and listen to conversations. As needed, ask guiding questions to help students plan. Sample questions:

  *Did you choose one person's idea or did you take pieces from more than one design?*

  *How did you make sure everyone had a voice?*

  *How will you attach the chair to the cable?*

  *What holds up the top end of the lift?*

- After each group agrees to a plan, each student in the group draws the final plan. Students then compare their ideas of the final plan to be sure they match. If they do not, they need to discuss them until they agree on one plan.

**Build**

*60 minutes*

- Tell students it is time to build the ski lifts. Call the Materials Managers up to collect the materials listed in the group's plan (two items from each of the four sets). Show them where they can get passengers (weights) for testing the ski lift as they build. Set the timer for 50 minutes.

- Remind students that they planned their designs as a guide to follow but that plans can and do change sometimes. If they need to change their plans, the Speaker must first get permission from the teacher, and then the Recorder needs to update the design. The Timekeeper must keep everyone on task, even when there's a change, because making a change does not allow the group more time.

- As students work, circulate and visit with each group. Praise students for good problem-solving efforts. Ask guiding questions to support critical-thinking skills. Sample questions:

  *Are you testing as you go or waiting until the end?*

  *Do you know if the ski lift will hold two passengers? How can you find out?*

  *Can the ski lift move the passengers in under 2 minutes? How can you find out?*

  *How can you make sure that your ski lift doesn't fall over?*

- Store finished ski lifts in a safe place until it is time for Test & Present.

**5** **Test & Present**

*15 minutes per group*

- Set up an area to test the ski lifts. Place a desk in an open space, measure 4 feet from the desk, and put a marker on the floor to show the starting place.

- Tell students that for this Challenge, a chair lift is defined as a ski lift that can carry from 2 to 4 people, and a gondola is defined as a lift that can carry a whole group of people. (There are other differences, but these simple definitions are useful for this Challenge.) Tell students that this test will show whether their creation works best as a chair lift or a gondola.

- Ask the first group to come to the test area with their ski lift. Invite students in other groups to ask questions before the test begins. Give the Recorder a Test Results sheet.

- Test the ski lift:

    - The **Materials Manager** kneels on the floor at the starting place.

    - The **Materials Manager** places two weights in the chair or gondola.

    - The **Timekeeper** sets the stopwatch to zero.

    - The **Materials Manager** uses the ski lift to raise the passengers from the floor to the top of the desk.

    - The **Timekeeper** measures how long it takes for the load to reach the top of the desk safely.

    - The **Recorder** writes down the time and number of weights used as passengers.

    - If the first test worked, the **Materials Manager** adds one more weight to the chair or gondola.

    - Repeat until the lift fails. Do not count data from the failed test.

- When all groups have tested their ski lifts, give each time to complete their Test Results sheet. If students do not yet know how to divide, then use the shortest time, rather than the average time, to determine if the ski lift meets the time criterion. Then give each student a Challenge Reflection sheet to complete.

## Opportunities for Differentiation

**To make it simpler:** In place of the Set 1 options, provide students with a plastic cup with one side cut out to look like a chair, and a front edge no higher than a half-inch. Tell students this is the chair for the lift. Rather than lift weights, the ski lift needs to raise a ping-pong ball without dropping it.

**To make it harder:** Use large marbles as the load, and require the ski lift to hold the load for the entire trip. Or, allow students to choose their own starting place, but require them to measure the height of the desk, the distance from the desk to the chosen starting place, or the angle from the floor to the slope. Have students represent these values on graph paper as a labeled triangle.

Names _____  _____

_____  _____

# Ski Lift Test Results

1. Does the ski lift have something to carry passengers?  _____

2. Is the ski lift human-powered?  _____

3. Test the number of passengers (weights) the lift can carry. Shade in boxes that are not used. Add the total number of seconds and divide by the total number of tests to get the average time.

| Test Number | Number of Passengers | Time (seconds) |
|:---:|:---:|:---:|
| 1 | 2 | |
| 2 | 3 | |
| 3 | 4 | |
| 4 | 5 | |
| 5 | 6 | |
| 6 | 7 | |
| 7 | 8 | |
| 8 | 9 | |
| 9 | 10 | |
| 10 | 11 | |

4. Average time _____ seconds

5. How many passengers does the lift carry? _____

6. Ski lift qualifies as a _____ (chair lift or gondola)

# Winter Coats

## Curriculum Connections

Science: adaptations, habitats, energy transfer
Mathematics: graphing

## Criteria for Product

- Must have an inner and an outer layer
- Must have stuffing sealed in between the two layers
- Should be thin, not bulky
- Should be attractive, as scored by classmates
- Must keep the wearer warm and dry for as long as possible

## Constraints for Challenge

- May use only materials provided in the kit, but may trade with other groups for more of the same materials
- Must complete each stage in the time allowed

### Challenge
Every winter people look for new ways to stay warm while still looking good. Students design, build, and test prototypes of winter coats.

## Materials

### Winter Coat Materials (for each group)

- 1 sandwich bag full of fiberfill stuffing
- 1 sandwich bag of beanbag pellets
- 1 sandwich bag full of feathers
- 1 × 8 × 16-in. piece of seat cushion foam
- 1 sq. ft. of quilt batting
- 1 sq. ft. of faux fur
- 1 sheet of felt
- 2 sheets of thick paper towels
- 1 sheet of construction paper
- 1 ft. of aluminum foil
- 1 ft. of waxed paper
- 1 small, white, plastic trash-can liner

- 1 shopping bag (to hold materials kit)

### Tools

- Metric rulers
- Scissors
- Staplers
- Masking tape
- Duct tape
- White glue
- Crayons
- Colorful permanent markers

### Additional Materials

- Books about polar animals and their adaptations (optional)
- Ice chest (teacher use)
- Shoebox-size plastic boxes to hold ice (1 for each group)

- 1 large bag of ice (or make your own)
- A few extra sandwich bags full of feathers
- A few extra pieces of faux fur
- 8 or more resealable plastic sandwich bags (without slider at top)
- 1 large container of shortening
- 8–10 manila file folders or cardstock
- Chart paper
- Markers
- Plain paper
- Pencils
- Digital timer

## Before You Begin

- Purchase a large bag of ice or make enough trays of ice for each group to have 6 to 8 cubes.
- Make a blubber layer for each group: Scoop shortening into a resealable bag and then seal the top. Carefully squish the shortening to form an even layer. Add enough shortening to make a full layer, but not so much that the bag is bursting at the seams.
- Make a winter coat template for each group: Sketch an outline of a simple winter coat that is about 4 inches by 6 inches. Trace the pattern onto a file folder and cut it out to make a template. Make one template per group and include it in the materials kit.
- For each group, make up one sandwich bag each of fiberfill, feathers, and beanbag filling.
- Assemble a materials kit for each group, following the Winter Coat materials list.
- Gather books, website links, and other resources about polar animal adaptations.
- Make a set of Job Cards (page 13) and a copy of the Test Results (page 89) for each group.
- Make a copy of the Challenge Reflection sheet (page 14) for each student.
- Set up the digital timer where students can check on it.
- Display the Criteria and Constraints where students can see them.

## 5-Step Process

**Investigate**
*Session 1*
*45 minutes*

- For each group, set up a plastic container with ice from a nearby ice chest.
- Divide the class into small groups of three to four students and provide each group with a container of ice.
- Invite students to take turns placing one hand on top of the ice and keeping it there for as long as possible. (Watch to be sure that no student keeps his hand on the ice long enough to injure it.) Encourage students to time each other using the wall clock.
- After all students have investigated the ice, ask for volunteers to describe how it felt. How long were they were able to keep their hands on the ice? Was getting cold the only problem or did they get wet, too?
- Ask students to identify habitats that include ice (polar regions, the Arctic and Antarctic). Next, ask what kinds of animals live in these habitats. Remind students that the features that help a living thing survive in its habitat are called *adaptations*. What kinds of adaptations do animals have for staying warm and dry in polar habitats, where it's icy and cold?
- Tell students that they'll be investigating how three adaptations, blubber, feathers, and fur, protect against ice. If students did not mention it when you asked about adaptations, explain that blubber is a layer of fat under the skin that keeps some polar animals warm.
- Demonstrate how to test the materials. Lay a blubber layer on one of the containers of ice and place your hand on the blubber layer.

- Distribute to each group one blubber layer, one piece of faux fur, and one sandwich bag of feathers. Instruct students to choose one person in the group to test each material to see how long she can keep her hand on the ice when it's protected by fur, feathers, or the blubber layer.
- Instruct students to switch materials around until all students in the group have had a chance to test all three materials.
- When students are finished, return the ice to the ice chest. Ask for volunteers to describe the difference between touching the ice directly and touching it through the blubber, fur, or feathers. Was one substance better than another for keeping their hands warm and dry?

**1 Investigate**
**Session 2**
**30 minutes**

- Remind students about the adaptations they explored in Session 1: fur, feathers, and blubber. Lead a discussion about how these adaptations help animals to survive in polar habitats. Share photos and other resources about polar animals and discuss how each animal is adapted to its habitat.
- Let students know that engineers sometimes study plants and animals to get ideas for designing objects used by humans. This practice is called *biomimicry*. Humans have used biomimicry to design winter clothes for many years. They think about adaptations that keep animals warm and dry, and come up with materials and designs inspired by those adaptations.
- Tell students that their next engineering Challenge is to design and make winter coats that would keep someone warm and dry. Of course, real winter coats are large, so they will work with small models. Explain that scientists and engineers also build small models, called *prototypes*, to test how well their ideas work.
- Show students the template they will use for their winter coat models. Explain that since they are testing the material rather than the design of the coat, they won't need to make a coat they can put on. Instead, they'll test their model coats on ice the same way they tested the animal adaptations.
- Review with students the Criteria and Constraints, as well as the time limits for Brainstorm, Plan, and Build.

**2 Brainstorm**
**30 minutes**

- Show students the kit materials. Their Challenge is to design and build a model coat that meets all the Criteria and uses only the materials available in the kit. Ask students:

  *What are some ways you keep warm and dry?*
  *What materials are used in items that keep us warm and dry?*
  *How do birds keep warm? How do deer keep warm?*
  *Can a material that keeps an animal warm also keep a human warm?*
  *What questions do you have about the Challenge?*

- Divide students into groups of three to four, each based on similar learning styles—kinesthetic, auditory, visual, or others.
- Tell students that for this Challenge they will brainstorm as a group. But each group member will still draw her own ideas.

- Show the Test Results sheet so students understand how their model coats will be judged against the criteria.
- Circulate through the room as students work, asking questions to prompt thinking. Suggested questions:

  *What are some properties that these materials have? How are these properties helpful for keeping someone warm or dry?*

  *How could you use these materials to make a winter coat?*

  *How could you layer the materials without making the coat too thick?*

  *Are any of these materials waterproof?*

- Once students are familiar with the materials, make sure they all have plain paper and pencils. Set the timer for 10 minutes. Have students think up as many designs as they can for a model coat that would keep a hand warm and dry on the ice. Explain that each person works on her own designs but may talk with other group members while working. Encourage students to label diagrams with the materials they plan to use.
- While groups brainstorm, listen in on their conversations. Ask questions to deepen the discussion and offer different points of view to spur new ideas. Suggested questions:

  *Is it easier or harder to brainstorm with a group (instead of by yourself)?*

  *What properties are important for the outer layers? Which materials have these properties?*

  *Which materials are best for the inside layer?*

### 3 ▸ Plan

*30 minutes*

- Review the responsibilities of each job as described on page 12: Speaker, Timekeeper, Materials Manager, and Recorder. Pass out Job Cards to each group and invite students to choose jobs. In groups of three, one student takes two jobs.
- Give each group's Recorder a sheet of chart paper. Tell students it's time to come up with one plan that they all agree to build. A good place to begin is to review all of their ideas from Brainstorm. As a group narrows down their choices, the Recorder draws ideas in pencil on the chart paper.
- If a group needs more of one material to build according to their plan, then the Speaker should find the Speakers in the other groups to discuss what they need, and what they can offer in trade. No trade should be made yet, just an agreement to make the trade.
- Circulate through the room, looking at plans. Ask questions to better understand student thinking and to push students to think about previous experiences that apply. Sample questions:

  *What experiences do you have with winter coats that can help you with your design?*

  *What experiences do you have with the materials in the kit? What are the properties of these materials?*

  *How do you know the coat will keep you warm?*

*Is everyone sharing ideas and working during planning?*

*How can you get more of a material if you need it for your design? Which group member should set that up?*

- Once the drawing is finished, the Recorder uses markers to color-code and label the materials the group is planning to use in their coat.

## Build
### 60 minutes

- Tell students it is time to build the coats. Set the timer for 60 minutes and have Materials Managers collect kits for their groups. Materials Managers should also make any trades that the Speakers arranged during Plan.

- Remind students that their plan is the basis for building. If the plan changes as they build, the Speaker notifies the teacher and the Recorder updates the plan using a different color.

- Continue to monitor work. Praise students and encourage teamwork. Ask guiding questions to develop ideas further, such as:

    *How is each person applying something he is good at to building the winter coat model?*

    *Why did you decide to use those materials?*

    *How are you making your coat attractive to a buyer?*

- Store finished coats in a safe place until it's time for Test & Present.

## Test & Present
### 10 minutes per group, then 10 minutes as a class

- Give a Test Results sheet to each Recorder. Tell students that all groups will test their model coats on the ice at the same time. But before testing, each group will present its coat to the group and survey the class about the coat's attractiveness.

- Call on the first group and set the timer for 5 minutes. Ask the Materials Manager to show the model coat while the Speaker explains why they chose the materials they did. Invite students from other groups to ask questions; allow up to three questions per group.

- Instruct the Materials Manager to measure the thickness of the coat. This is done by laying the coat on a desk or table and using a ruler to measure its greatest thickness from top to bottom. The Recorder writes the measurement on the Test Results sheet.

- Tell students they will now vote on the attractiveness of this model coat. Remind students to be respectful but honest. They will rate the attractiveness of the coat on an A, B, C scale, with A as "Awesome," B as "Very nice," and C as "Okay." Each student may vote only once.

- Instruct the Speaker to call out the score choices (A, B, C) one at a time. For each number, the Recorder counts the number of raised hands and records it in the tally chart on the Test Results sheet.

- Continue this procedure until all groups have presented their model coats, measured their thickness, and tallied attractiveness scores.

- At the end of the presentations, tell students that all groups will now test how long their coats keep the wearer warm and dry in icy conditions. Provide each group with a container of ice, as in Investigate, and get ready to start a timer that all Timekeepers can see. Instruct the Materials

Manager to place the coat on top of the ice and the Timekeeper to get ready to watch. When you say, "Go," each Speaker places one hand on top of the coat. Let the Speakers know that when their hands get too cold or too wet, they should say, "Stop," and remove them from the coat. At that exact moment, the Timekeeper notes the time in minutes and seconds, and the Recorder adds this information to the Test Results sheet. The Speaker should also tell the Recorder whether she stopped because her hand was cold, wet, or both.

- If your students understand the mathematical concepts of mean and average, have them find these values for their coats' attractiveness scores. Give groups a few minutes to graph the attractiveness scores.

- Discuss the recorded results to determine which criteria each group would work to improve, if given time. While reviewing the results, ask students:

  *Which coat was the most attractive (according to the voting)?*

  *Which coat kept the wearer dry the longest?*

  *Why do you think that coat worked the best?*

  *Were some stuffing materials warmer than others?*

  *Was the thickest coat also the warmest?*

  *Did thickness have anything to do with its attractiveness?*

  *When would it make sense to choose a more attractive coat over a warmer one, or the other way around?*

  *Why might a warm, thin coat be more appealing than a warm, bulky one?*

- Give each student a Challenge Reflection sheet to complete.

## *Opportunities for Differentiation*

**To make it simpler:** Provide students with prebuilt outer and inner layers, already attached. Students then decide which materials to use as filling.

**To make it harder:** Rather than gluing or stapling the coats shut, provide students with blunt needles and yarn. Instruct students to sew the layers shut so that no stuffing pokes out. Or, require students to make a two-sided coat or mitten that they can slip their hands inside.

Names _____   _____

_____   _____

# Winter Coat Test Results

1. Does the coat have both an inner and outer layer? _____

2. Is the stuffing sealed between the two layers? _____

3. The coat is _____ centimeters thick.

4. Tally Chart

| Attractiveness Score | Number of Votes |
|---|---|
| A.  Awesome | |
| B.  Very Nice | |
| C.  Okay | |

Graph of Attractiveness Scores

| | | | |
|---|---|---|---|
| > 10 | | | |
| 10 | | | |
| 8 | | | |
| 6 | | | |
| 4 | | | |
| 2 | | | |
| Number of Votes | A. Awesome | B. Very Nice | C. Okay |

5. The coat kept the wearer's hand warm and dry for _____ seconds.

6. Was the wearer's hand cold, wet, or both at the end? _____

# Greenhouses

## Curriculum Connection

Science: plant needs, photosynthesis, water cycle

## Criteria for Product

- Must be portable and easy to set up
- Must stand by itself
- Must open, close, and seal
- Plant should need watering no more than once in 10 days
- Must use exactly 10 items: for example, 1 cereal box, 2 cardboard tubes, 1 bag, 3 rubber bands, 2 craft sticks, and 1 sheet of paper

## Constraints for Challenge

- Must use only the materials provided
- Must be something built by the group (not something the group used without changing it)
- Must complete each stage in the time allowed

### Challenge

Whether it's cold temperatures or lack of fertile soil, it can be difficult to grow vegetables outdoors in some places. Students design and build portable indoor greenhouses.

## Materials

### Greenhouse Materials (for the class)

- Assorted large cans
- Assorted cardboard boxes
- Assorted plastic bottles
- Assorted plastic tubs
- Assorted paper bags
- Newspapers
- 2 large rolls of plastic wrap
- 50 pipe cleaners
- 2 rolls of paper towels
- 100 paper plates
- 200 craft sticks
- 25 wire clothes hangers

- Paper towel and bath tissue tubes
- A few rolls of clear tape
- A few rolls of masking tape

### Tools

- Pointed scissors
- Rulers or tape measures
- Penknife (teacher use only)

### Additional Materials

- Paper towels, 1 sheet for each student
- Resealable sandwich bags, 1 for each student

- Popcorn kernels, 3 for each student
- Permanent markers
- Chart paper
- Markers
- Plain paper
- Pencils
- Digital timer
- Paper cups, 8- or 10-oz. size, 1 for each student
- 1 small bag of potting soil
- Digital video camera (optional)

## Before You Begin

- Contact families to ask for donations of clean recyclables.
- Make a set of Job Cards (page 13), a copy of the Greenhouse Presentation Directions (page 96), a copy of the Observations (page 95), and a copy of the Test Results (page 97) for each group.
- Make a copy of the Challenge Reflection sheet (page 14) for each student.
- Set up the digital timer where students can check on it.
- Display the Criteria and Constraints where students can see them.

Get Set

# 5-Step Process

**1** ⟩ *Investigate*
*Session 1*
*45 minutes*

- Let students know that they'll be growing popcorn plants to prepare for their next engineering Challenge. Provide each student with a paper towel, resealable sandwich bag, and three kernels of popcorn. Have them fold their paper towels so they will fit inside their plastic bags.
- Instruct students to wet their paper towels, place their popcorn kernels inside the folded paper towels, put the paper towels inside their plastic bags, and seal the bags.
- Have students use permanent markers to write their names on their sandwich bags. Put the sandwich bags in a sunny place—for example, tape them to a window.
- Tell students they have just made miniature greenhouses to get their popcorn plants started. Greenhouses provide plants with conditions that support plant growth. They are especially useful in areas with short growing seasons. Ask students:

    *What do plants need to grow and thrive? (As needed, review the list in your curriculum.)*

    *What conditions does a greenhouse create for the plants?*

    *Why do you think you used plastic bags instead of fabric or paper?*

    *Does the greenhouse material need to be clear? Why or why not?*

    *Why did you dampen the paper towels?*

    *How can you tell if a greenhouse is working? What will you see?*

- Explain that water is always evaporating, and the warmer the air, the faster it evaporates. When the sun is out, the air in a greenhouse gets warmer than the air outside the greenhouse because the air cannot escape. The water that evaporates inside the greenhouse cannot escape, either. So when that water vapor reaches the greenhouse wall (the inside surface of the bag), it is cooled and turns back into liquid water drops. This liquid is called condensation.
- Remind students to look for condensation when they check on their plants.

**1** *Investigate*

*Session 2*

*30 minutes*

- Set the items from the Greenhouse Materials list out on a table.
- Let students know that once the corn plants grow larger, they will need a larger greenhouse. So their next Challenge is to design and build a portable greenhouse.
- Show students the table full of available materials. Ask students:

  *What are some properties that these materials have?*

  *How could you use these materials to make a greenhouse?*

  *How could you change the shape of some of these materials to make them work for you?*

  *Which materials would work to hold up the greenhouse?*

- Invite students to offer suggestions for ways to sort the materials into categories for this Challenge. Guide them to think about categories such as wall materials, support materials, roof materials, door materials, water-holder materials, and so forth. Be sure they understand that a material may fall into multiple categories and that there are no right or wrong answers.
- Review with students the Criteria and Constraints, as well as the time limits for Brainstorm, Plan, and Build.

**2** *Brainstorm*

*20 minutes*

- Provide students with sheets of plain paper. Tell students to fold their papers to make four sections and write Brainstorm across the top. Have them work alone quietly for 15 to 20 minutes to design four different greenhouses, drawing one in each section.
- Remind students to count the items they use in each design to be sure they have 10. Ask them to label each item.
- As students work, move through the room, looking at designs. Ask guiding questions to encourage students who are rushing to focus on details, and to help students who are "stuck" to start coming up with ideas. Suggested questions:

  *Which material will you use for that part of the greenhouse? What makes that a good choice?*

  *How will this design help the plant get light for photosynthesis?*

  *How does this design help the plant get enough water?*

- As students finish, have them review their designs and circle the one they like best.

**3** *Plan*

*30 minutes*

- Tell the class that they will be allowed to form their own work groups. Remind them that a group should have a mix of people with different learning styles—that is, a mix of people who are good at doing different things.
- Invite students to group themselves in threes and fours. Assist students who are having difficulty finding a group, or who are trying to form a group that is too large.

- Pass out Job Cards to each group and review the responsibilities of each one (page 12). Explain to students that for this Challenge, it is up to the group to decide how to choose or assign jobs. Walk around the room. Listen to student conversations about jobs but do not interject. Allow students to work out issues on their own.

- After students have chosen their jobs, tell them it is time to come up with one group plan. Have students place their Brainstorm sheets on the table where everyone in the group can see all the designs. Remind students to take turns sharing ideas and to let everyone have a voice in the decision making.

- Set the timer for 20 minutes. Check in with each group and ask guiding questions to help students plan and extend their thinking. Sample questions:

    *How did you decide on the plan? Where did each idea come from?*

    *Have you checked to be sure you have 10 items?*

    *Have you started thinking about who will do each part when building the greenhouse?*

    *How does your greenhouse help condensation to form?*

    *What are you doing to keep water in the greenhouse without refilling often?*

- After a group agrees to one design plan, each student in the group draws her own version of it. All plans in a group should look alike. If they do not, students need to discuss where their plans differ and then revise until they have one group plan.

## 4 Build

*50 minutes*

- Tell students they will now collaborate (work together) to build their greenhouses. Explain that they should choose one plant out of all the plants in their bags to be the "greenhouse plant."

- Set the timer for 50 minutes. Invite Materials Managers to gather materials from the table.

- Remind students to follow the group plan for as long as it works for them. When the plan no longer works, they need to agree on how it should change. Then the Speaker gets approval for the change from the teacher and the Recorder updates the plan, making changes in a different color or drawing up a new Plan 2.

- Remind Timekeepers to be sure everyone's efforts are coordinated so the group finishes building before the time is up.

- Encourage students to share responsibilities and work as a group. Ask guiding questions to support critical-thinking skills. Sample questions:

    *What is each person's responsibility as you build?*

    *Are you testing how it opens and closes and how well it stands up?*

    *How did you choose which plant to use?*

- Instruct the Timekeeper to take the chosen corn plant out of the plastic bag and plant it in a cup of soil. Once planted, she should measure the height of the plant from the soil to the top and record it.

- You may choose to have the other students in each group plant their corn plants in cups as well. Then, they may either take home the plants not used in the greenhouse, or keep the plants in school and compare the growth of the plants outside the greenhouse to the growth of the plant inside the greenhouse. Explain to students that their plants should not, however, "take turns" in the greenhouse, because the idea is to observe how well one plant grows in the greenhouse conditions.
- Have students place the "greenhouse plant" inside its finished greenhouse. Place the greenhouses along the window until it is time for Test & Present. (For classrooms without windows, determine an appropriate location in the building to place greenhouses where they will get enough light, be accessible for students to observe, and not be disturbed.)
- Distribute an Observations sheet to each group. Instruct the Recorder to complete the first row of Observations.
- Tell students they need to observe the greenhouses and complete the Observations sheet about every other day for two weeks.

## 5  Test & Present

*10 minutes per group*

- Begin Test & Present at the end of two weeks, when students have completed their Observations sheets.
- Give each group a copy of the Greenhouse Presentation Directions.
- Tell students to be ready to ask questions during these Challenge presentations.
- Ask the first group to bring their greenhouse, with its plant, to the front of the room. Give the Recorder a Test Results sheet.
- Allow the Speaker to direct the rest of the group, using the Presentation Directions. The Materials Manager sets up the greenhouse, the Timekeeper measures the plant, and the Recorder writes down the results.
- Invite students to ask questions. Allow five questions per group.
- Distribute a Challenge Reflection sheet to every student when all groups have finished.

## Opportunities for Differentiation

**To make it simpler:** Provide students with a basic design for the greenhouse that does not have a stand or a door. Allow students to modify the basic design so it meets the product criteria. Or allow them to use materials other than the ones on the list.

**To make it harder:** Add a requirement to the criteria that the greenhouse must be a certain length and width or must be able to hold multiple plants.

Names _____     _____

             _____     _____

# Greenhouse Observations

Observe the plant in the greenhouse once every other day for 10 days.
Record observations below.

| Today's Date | How many days since you refilled the water? | Did you need to refill the water today? (Y/N) | How is the soil today? (Dry, Damp, or Wet) | How tall is the plant in inches? |
|---|---|---|---|---|
| | 0 (setup day) | | | |
| | | | | |
| | | | | |
| | | | | |
| | | | | |
| | | | | |

How much did the plant grow, in inches, from the first to the last observation?

Last height _____ − First height _____ = _____ inches

# Greenhouse Presentation Directions

1. Point out each of the 10 materials used.

2. Show how the greenhouse is portable.

3. Set it up and show how it stands on its own.

4. Show how it opens, closes, and seals.

5. Report how many times the plant had to be watered in 10 days.

6. Report how much the plant grew while in the greenhouse.

Names _____    _____

_____    _____

# Greenhouse Test Results

1. List the 10 materials you used.

    _____    _____

    _____    _____

    _____    _____

    _____    _____

    _____    _____

2. Is the greenhouse portable and easy to set up? _____

3. Can the greenhouse stand by itself? _____

4. Does your greenhouse open, close, and seal? _____

5. How many times did you need to refill the water? _____

6. Did you need to refill it more than once in 10 days? _____

7. Was there condensation most times you checked? _____

8. Was the soil damp (not wet or dry) most days? _____

9. Was your greenhouse successful? _____

10. Explain why you think your greenhouse was or was not successful.

    _____

    _____

# Bird Feeders

## Curriculum Connection

Science: adaptations, needs of living things

## Criteria for Product

- Birds are able to get to the feeder easily
- Birds are able to get at the food easily
- Feeder keeps the food safe from animals other than birds
- Made of materials that are safe for all animals

## Constraints for Challenge

- Exactly 7 items from the materials list must be used
- Must complete each stage in the time allowed

### Challenge

Winter conditions challenge animals, and they may have difficulty meeting their basic needs for survival. Students design and build bird feeders.

## Materials

### Bird Feeder Materials (for the class)

- To be determined during class Brainstorm
- Alternatively, work from the Ideas list (page 104)

### Tools

- Ruler
- Scissors
- Hole punches

### Additional Materials

- Computer with Internet access (optional)
- Projector for computer (optional)
- Chart paper
- Markers
- 1 plastic spoon
- 1 paper plate
- 1 paper cup
- 1 set of chopsticks
- 1 plastic fork
- Duct tape or masking tape
- Digital timer
- Bag, jar, or other container (to hold paper slips)

# Before You Begin

- Locate digital and print resources about birds local to your region. Be sure the resources contain good pictures of the birds. Include different birds that eat the kinds of foods you are planning to use for the feeders. If you already know the names of some birds that live in your region, a good reference site is All About Birds maintained by the Cornell Lab of Ornithology (www.allaboutbirds.org). (The site is not set up to identify a bird based on its appearance, nor does it provide lists of birds that are common in certain areas.)

- Decide whether you want to brainstorm a class list before getting materials, or if you prefer to gather materials in advance from the Ideas list (page 104). Either way, arrange to contact families for donations of recyclables, other materials, and foods, or plan to gather materials yourself.

- Choose a location on the school grounds for the finished bird feeders.

- Figure out how many groups of four students are in your class, and how many of three are left over. Write each group number on slips of paper, so there is one slip for each student. For example, 23 students make five groups of 4 and one of 3. Make four slips with number 1, four more with number 2, and so forth through number 5, but only three slips with number 6. Place all slips in a bag, jar, or other container.

- Make a set of Job Cards (page 13) for each group.

- Make a copy of the Presentation Questions (page 105) and Product Information (page 106) for each group.

- Make a copy of the Challenge Reflection sheet (page 14) for each student.

- Set up the digital timer where students can check on it.

- Display the Criteria and Constraints where students can see them.

# 5-Step Process

## 1 Investigate
### Session 1
### 40 minutes

- Display images of local birds. Invite students to discuss anything they know about these birds, such as whether they've seen them before, and if so, what they were doing.

- Tell students that although all these birds live in your area, they have different habits. For example, different birds eat different foods. Explain that in many cases, a bird's beak is related to the foods it eats. Humans find it convenient to eat different foods with different utensils. A bird's beak is its eating utensil.

- Set out the plate, cup, spoon, fork, and chopsticks. Pour a little birdseed on the plate. Tell students that the utensils stand for different birds' beaks. Invite students to take turns using the utensils to pick up birdseed and place it in the cup. Instruct students to watch what happens as each utensil is tested. Ask students:

  *Would it be easier to get at the seeds if they were in a different container?*

  *What kind of container might make it easier? Why?*

  *What are some everyday items that have a good shape?*

- Direct students to look at the pictures of local birds again and compare them to the utensils. Ask students:

    *Which of these utensils are most like the beaks that your local birds have?*

    *How do you think most of the local birds get their food?*

    *What kinds of foods, other than seeds, might be good for these birds?*

- Mention to students that today's activity will help them in their next engineering Challenge.

**1**

*Investigate*
*Session 2*
*30 minutes*

- Remind students of the activity in which they compared eating utensils to birds' beaks. Tell students that they have an important job to do. The local birds need food, so they will build bird feeders.

- As a class, discuss some different foods that students think local birds would eat. Explain that some foods humans eat may not be good for birds, even if birds will eat them. So we need to be careful about what we feed birds. With this in mind, share with students the list of bird foods from the Ideas sheet (page 104). Let them know that the foods on this list are the only ones allowed in their bird feeders.

- Instruct students to think about what they learned about local birds and their beaks. Ask students:

    *Which foods on the list would be easiest for local birds to eat?*

    *What can you use to hold the food where birds can reach it?*

    *Do all bird feeders need to be containers, or can bird foods be stuck to something?*

    *What other animals might take the food?*

    *Can you design a bird feeder that lets birds get the food but keeps it safe from other kinds of animals?*

    *Do you have any questions about this Challenge?*

- Tell the class that they'll come up with a class list of materials they could use to build bird feeders. Suggest students keep in mind the foods they'll be using. They should also think of materials that are as natural as possible, or that will break down in the environment. Natural materials ensure that animals are safe and that the environment is not disrupted more than necessary.

- Write a list of possible materials as students answer the following questions. Ask students:

    *What are some materials that could hold the food?*

    *Will you need any materials to keep the feeder in place?*

    *How will birds hold onto the feeder while they eat? What materials do you need for a ledge or perch?*

- Alternatively, you may assign students to bring in bird feeder materials similar to the ones on the Ideas sheet (page 104) instead of creating a class list.

- Review with students the Criteria and Constraints, as well as the time limits for Brainstorm, Plan, and Build.

**2 Brainstorm**

*20 minutes*

- Display the bird foods and the building materials on a table. As a class, briefly discuss which materials could be used to make bird feeders for each of the foods that are available.

- Tell students they will work in small groups to brainstorm bird feeder ideas. To form groups, students draw numbers out of the bag, jar, or container you prepared earlier. Students with the same number are in the same group.

- Provide each group with a sheet of chart paper and four markers. Tell them that they have 15 minutes to brainstorm ideas for a bird feeder. Ask students:

    *Which food(s) do you want to provide the birds?*

    *How will the feeder hold the food(s)?*

    *How could you use each of these materials?*

    *Where will you place the feeder?*

    *How can you keep the feeder safe from animals other than birds?*

- Ask students to talk in their groups and offer suggestions. Whoever comes up with an idea may draw it on the chart paper. Remind students that they are not planning yet. Encourage them to be "messy" with ideas, words, and drawings. This is not the time to choose or discard any designs. Instead, come up with the next idea.

- Remind students that each bird feeder design needs to include 7 items (food items do not count toward the 7). They should brainstorm as many different designs as they can in 15 minutes, each with a list of 7 items. As needed, point out that the same item may appear on more than one list.

- Walk from group to group, looking at each chart paper and asking guiding questions that push students beyond their initial ideas. Suggested questions:

    *What kind of birds will use this feeder?*

    *How will birds get at the food?*

    *How will you use that material in your bird feeder?*

**3 Plan**

*30 minutes*

- Distribute Job Card sets to each group and have students choose jobs by drawing a card at random. Remind students about the responsibilities of each job (page 12).

- Tell students that it's time to review their Brainstorm results and, as a group, create one plan that meets all the Criteria. Set the timer for 20 minutes.

- Walk around the room, facilitating the work. Ask guiding questions to help students work together to design the device. Sample questions:

    *How will birds perch on the feeder while eating?*

    *Which bird are you thinking of, for this feeder? What is its beak like?*

    *How does this feeder hold the food you plan to use?*

    *Does everyone agree on this design? If not, which parts do you disagree with? Why?*

*Who thought of this part? What was your thinking when you designed this part?*

- After 20 minutes, or when a group has reached a decision, each student draws a diagram of the final plan on a separate sheet of paper. All plans within a group must look the same. If they do not, then let the group know that they've had some miscommunication. They need to review their plans until everyone understands and agrees.

## Build
### 60 minutes

- Tell students that it's time to build the bird feeders. Set the timer for 50 minutes.
- Materials Managers collect all the necessary items. Only materials listed in the group's plan may be used.
- Remind students that the plan is like their blueprint. All good builders follow the blueprint. Tell them that following the blueprint doesn't always work the first time, so they may need to adjust it. If they do, then the Speaker gets approval from the teacher and the Recorder changes the plan so it shows the actual finished product. Recorders should make changes in a different color or draw a new Plan 2.
- Instruct the Timekeeper to watch the timer and update the group on how much time is left. Remind groups that the Timekeeper is in charge of keeping the task moving and making sure everything is finished on time.
- Visit each group to see how things are progressing. Sample questions:

    *Is everyone helping to build the bird feeder?*

    *What do you know about birds' beaks? How did you use that information in your design?*

    *Will everything hold together when the food is added, and when birds land on it?*

    *What is the hardest part of building this feeder?*

    *How does your design help keep the food safe from raccoons and squirrels?*

    *Where will you place your bird feeder? How will it stay up?*

    *How do you know that birds will be able to get the food?*

- Store finished bird feeders in a safe place until it's time for Test & Present. (If feeders have food already attached, be sure to close them up in airtight containers to keep away ants and other animals.)

## Test & Present
### 15 minutes per group

- Lead the class outdoors to the site you chose earlier. Bring the bird feeders, the Product Information sheets, Presentation Questions, and pencils.
- Have groups take turns presenting. Give each Speaker a copy of the Presentation Questions. Give each Recorder a Product Information sheet and a pencil.
- Have the first group's Speaker read the "Before Setup" questions from the Presentation Questions sheet, or you may read them to the Speaker. Have the Speaker answer each question aloud.

- Instruct the Materials Manager to set up the group's bird feeder where it will be used by birds.
- Tell the Speaker to read and answer the "During Setup" questions.
- Invite the other students to observe how the bird feeder meets the criteria. Prompt the Recorder to write down their responses on the Product Information sheet.
- Prompt the Speaker to read and answer the "After Setup" questions.
- Invite students from other groups to ask any questions they have. Limit additional questions to three.
- Repeat until all groups have presented and placed their feeders.
- Distribute a Challenge Reflection sheet to each student.

## Opportunities for Differentiation

**To make it simpler:** Allow students to research homemade bird feeders, choose one, and create it with a slight modification. Internet searches for "bird feeder for kids" and "easy bird feeders" return good results.

**To make it harder:** Limit the number of different materials students have available to use. Add a requirement that birds must be able to reuse the feeder materials for building nests. Add follow-up observations over the next few weeks and record how many birds are seen feeding at each feeder each day.

# Bird Feeder Materials Ideas

For reference during Brainstorm and to use as a shopping list

### Bird Foods

- Birdseed
- Popped popcorn
- Fruit slices
- Peanuts
- Raisins
- Melon seeds
- Suet
- Sugar water

### Materials for Bird Feeders

- Paper towel and bath tissue tubes
- Soda and water bottles
- Milk cartons
- Cardboard boxes
- Yarn or string
- Newspapers

### Natural Materials for Bird Feeders

- Pinecones
- Sticks and twigs
- Coconuts and shells
- Citrus fruit skins

Note: *Substitute readily available materials if suggested items are not on hand.*

# Bird Feeder Presentation Questions

Read each question. Then say the answer for your group.

### Before Setup

1. How does our bird feeder work?

2. Did we follow our plan or did we make changes?

3. What changes did we make?

4. How well did our group work together?

### During Setup

5. Is it working the way we planned?

6. Can we set it up easily?

7. Are there changes we would make if we could?

### After Setup

8. Are we happy with our design?

9. How could we improve our design?

Names _____    _____

_____    _____

# Bird Feeder Product Information

1. How can birds get to the feeder?

   _____

   _____

2. How can birds get at the food?

   _____

   _____

3. How does this feeder keep the food safe from animals other than birds?

   _____

   _____

| Materials used (exactly 7) | This material is safe for animals (yes / no) |
|---|---|
| 1. | |
| 2. | |
| 3. | |
| 4. | |
| 5. | |
| 6. | |
| 7. | |

# Parachutes

## Curriculum Connections

Science: properties of air, gravity
Mathematics: graphing

## Criteria for Product

- Must take longer to reach the ground than the control load takes
- Must hold together when landing
- Must still be attached to the load after landing

## Constraints for Challenge

- Must use only the materials provided
- Must stay within a budget of $5.00
- Must complete each stage in the time allowed

### Challenge
From dropping supplies from a plane to landing a rover on Mars, engineers have spent years figuring out how to safely land objects on a surface. Students design and build parachutes to safely land a cargo load.

## Materials

### Parachute Materials (for the class)

- 1 box of facial tissues
- 1 roll of waxed paper
- 1 roll of aluminum foil
- 1 roll of paper towels
- 10 plastic shopping bags
- 25–50 lunch-size paper bags
- Plain paper
- Construction paper
- 50–100 craft sticks
- 50–100 straws
- 50 coffee filters
- A few rolls of masking tape
- A few rolls of clear tape
- 1 roll of string

### Tools

- Rulers
- Tape measures
- Scissors
- Hole punches

### Additional Materials

- Chart paper
- Markers
- Pencils
- Plain paper
- Bag, hat, or other container (to hold slips of paper)
- Digital timer
- Pretend money, $5.00 for each group, in an envelope or bag
- Box for the shopkeeper's "cash register"
- 1 balloon

- 1 empty plastic soda bottle (any size)
- 1 turkey baster
- Items to use as loads for the parachutes (identical plastic figurines work well)
- Computer with Internet access (optional)
- Projector for computer (optional)
- Stopwatches, 1 for each group
- Digital cameras, 1 for every 2 groups (optional)
- Computers with slideshow software (student use, optional)

**Note:** *Quantities are approximate and may be adjusted.*

# Before You Begin

- Recruit one or more adult helpers for the day that parachutes will be tested.
- Select a safe outdoor location for testing parachutes. For example, students could drop the parachutes out of a second- or third-story window onto a safe sidewalk outdoors.
- Figure out how many groups of four students are in your class and if there will be any groups of three. For example, 18 students make three groups of four and two of three. Make four slips of paper with number 1, four slips with number 2, and four slips with number 3, but only three slips with number 4 and three slips with number 5. Place all slips in a bag, hat, or other container.
- Make a set of Job Cards (page 13), a copy of the Price List (page 113), and a copy of the Test Results (page 115) for each group.
- Place parachute materials on a table. Make signs for the price of each material or place a copy of the Price List on the table.
- Locate images of parachutes of different shapes and sizes. An Internet image search with the word "parachute" offers several options.
- Make a copy of the Challenge Reflection sheet (page 14) for each student.
- For each group, place $5.00 in an envelope or bag.
- Set up the digital timer where students can check on it.
- Display the Criteria and Constraints where students can see them.

# 5-Step Process

## 1 Investigate
### Session 1
### 20 minutes

- Review or introduce to students the idea that air is a kind of matter, like water or soil or any other thing. Because it is matter, air has weight and takes up space, even though you can't see it. And because air is matter, it interacts with other things—other things can push air around and air can push back.

- Ask students to think for a minute about different ways that they can show that air takes up space. Discuss a few of the students' ideas as a class.

- Show students a balloon, plastic shopping bag, paper lunch bag, an empty plastic drink bottle, and a turkey baster. Ask a few students to come up, one at a time, choose one item, and use it to demonstrate that air takes up space. For example, a student might fill a bag with air and twist it closed to show that the air takes up space in the bag.

- Next, ask students to think about ways to show that air interacts with other things.

- Invite a few students to come up, one at a time, and use the same materials to demonstrate this. For example, a student might use the turkey baster to blow air at a wad of paper. The baster pushes the air and the air pushes the wad of paper.

**1** *Investigate*
*Session 2*
*30 minutes*

- Remind students that in the previous session they learned about moving air pushing on objects. Explain that air pushes on objects when the object is moving, not just when the air is moving. This phenomenon is called *air resistance* or *drag*. Ask students to think of times that engineers might need to think about air resistance—for example, when designing a car.

- Tell students that engineers use air resistance when designing parachutes, and that they will design parachutes as their Challenge.

- Explain that engineers often use computer simulations to test designs before building them. Computer simulations are based on real data, so they behave in a similar way to real test models. Students will use a computer simulation to design and test a parachute before designing and building their real parachutes.

- Use a computer projector or allow students to work on individual computers. Go to the website (www.pbs.org/wgbh/nova), search for "Design a Mars Parachute," and launch this interactive. In this simulation, the user designs a virtual parachute that safely lands a rover on the surface of Mars. Guide students through the Introduction, Requirements, and Design & Test, defining terms and paraphrasing the explanations as you go. Then allow free exploration of the Design & Test portion.

- As students are working, ask:

    *How does the canopy diameter (width) affect the way the parachute works?*
    *How does the width of the band affect the parachute?*
    *Why is it important to consider how thick the material is?*
    *How is the load attached to the parachute?*

- If time permits, show students images of parachutes that have different shapes and sizes. Ask students to think about how the shape, size, number of strings, and other variables might affect a parachute's performance.

- Allow students to touch and observe the materials they have available to build a parachute, so they can base their decisions on the materials' properties.

- Review with students the Criteria and Constraints, as well as the time limits for Brainstorm, Plan, and Build.

**2** *Brainstorm*
*15 minutes*

- Show students the "load" (plastic figurine) that their parachute needs to drop safely. Explain that they will brainstorm alone, but they will work in groups to plan and build the parachutes.

- Give each student a sheet of plain paper. Tell students that they have 10 minutes to come up with possible parachute designs that use the materials available.

- Encourage students to think about parachute size, shape, and materials. As needed, point out that it is more important to think about how the parachute works than what it looks like.

- Remind students that designing can be messy work. It is important for them to sketch as many ideas as possible—not to make one perfect drawing.

- Walk around the room, asking guiding questions. Suggested questions:

   *How long will that part be? How wide?*

   *How is the shape you've drawn like other things you've seen or used?*

   *How is the figure attached to the parachute in that design?*

   *Why did you choose that material? What are its properties?*

   *Can you think of another shape for a parachute?*

- After 10 minutes, ask students to circle their favorite or best design.

## 3 Plan
### 40 minutes

- Take out the container of group numbers that you set up before the Challenge. Let each student draw a group number at random.

- Have students get into their groups. Review the responsibilities associated with each job (page 12). Hand out the Job Card sets and ask students to choose jobs within their groups in any way they wish.

- Instruct students to take turns sharing their parachute designs from Brainstorm. They need to discuss how to combine their ideas into one design that the group will build together. Give a copy of the Price List to each group, and remind students they must stay within budget.

- Set the timer for 30 minutes. Check in with each group, complimenting good collaboration and asking guiding questions to help students stay on task and improve their designs. Sample questions:

   *How did you decide on this part of the design?*

   *Will you need to change something else about the parachute because of this change? Why?*

   *Do you know how big this parachute will be?*

   *How much of this material do you need? How can you be sure?*

   *Are your materials within budget?*

- As groups finish, hand out plain paper. Each group member draws the group's plan and writes a materials list. All plans within a group should match. If they do not, then the group needs to talk together until they all understand and agree to the same plan.

## 4 Build
### 45 minutes

- Decide whether you will allow groups to go over budget if any material needs replacing as they build. Let students know in advance what your policy will be.

- Tell students it's time to build the parachutes. Set the timer for 45 minutes.

- Give an envelope with $5.00 in pretend money to each Materials Manager. Invite them to come to the table to shop for the materials in their group's list.

- Remind students to refer to their group's plan for the entire build. If something in the plan will not work, then they may change the plan. But the Speaker must first get the change approved by the teacher, and the Recorder needs to show the change so that the plan reflects the finished product.

- Keep up with the progress of each group. Get involved in discussions but do not offer suggestions. Encourage students to let everyone help build the parachute. Ask guiding questions that remind students to use their science knowledge. Sample questions:

  *How does air push against things? How do other things push against air?*

  *Have the materials worked the way you expected so far?*

  *What are you doing to make sure your parachute holds together?*

  *Have you tested your parachute yet? How did you do it?*

  *Have you had any unexpected challenges yet? What did you do to solve them?*

- Store finished parachutes in a safe place until it's time for Test & Present.

## 5 Test & Present

*20 minutes to test; 15 minutes to graph; 30 minutes for slideshow prep; 15 minutes to present*

- Have your adult helper(s) join you, and hand her an extra figurine (load) that does not have a parachute. Explain that this is the control load, which will be dropped before any of the parachutes.

- Decide whether to allow groups to retest in the case of mishaps. If you wish to do so, appoint a pair of students as runners to carry parachutes back indoors and upstairs.

- While still in the classroom, have each group join with another group and give both groups one digital camera to share. Tell students that each group needs three pictures: Picture 1 is the whole group, with the Materials Manager holding the parachute. Picture 2 is the parachute against a blank background, showing each part. Picture 3 is taken during the test.

- Have the Materials Managers take their parachutes and go with the adult helper to the top of the drop site. Take the rest of the students to the bottom of the drop site, bringing stopwatches, Observations sheets, pencils, and digital cameras.

- The first test is finding out how long it takes the control load to fall. Instruct the adult helper to call out when she is ready to drop the control load. When she does, the Timekeepers can start their stopwatches. All Timekeepers measure the time it takes. (It will be very fast.) After the control load has dropped, choose the most common time measurement and instruct all Recorders to write it on their Observations sheets.

- Tell students that each group will take a turn, following this procedure:

  - The **Materials Manager** drops the parachute.
  - The **Timekeeper** measures how long it takes to drop.
  - The **Recorder** writes the results for all groups on the Observations sheet.
  - The **Speaker** takes a picture of the parachute dropping, and then picks up the parachute from the ground.

- Call each group in turn. Remind Timekeepers to measure times for all parachutes, not just their own. Record the most common measurement for each drop on the Observations sheets.

- Back indoors, have each group answer questions 1 through 3 on the Observations sheet. Then distribute a Test Results sheet to each group. Point out that the bar graph needs a scale on the left side. Explain that you need to choose a scale that works for the data of this particular test. The scale starts at zero and goes up in even numbers (or even increments) to a number that's just above the greatest measurement. Discuss the range in time measurements for all drops, and then choose the numbers to use for the scale. Have each group complete the graph.

- If you plan to do slideshows, help students upload photos from the cameras to the computers. Otherwise, plan for live group presentations.

- Help students create slideshows of their group's parachute demonstration, using presentation software or an online photo tool. Their slideshows should include captions, titles, or recorded narration.

- Have groups take turns presenting their results. The Speaker should announce why the group thinks their parachute performed the way it did.

- Invite students to ask questions. Limit the number of questions per group as needed.

- When all groups have presented, have each group answer questions 4 through 6 on the Observations sheet. Have each student complete a Challenge Reflection sheet.

## *Opportunities for Differentiation*

**To make it simpler:** Provide each student with a square paper towel, 4 strips of clear tape, 4 pieces of string cut to the same length, and a large paper clip. Guide students to create a parachute by taping each string to each corner of the paper towel, bringing the strings together at the bottom and securing them to the paper clip (the load). Allow students to explore with these parachutes to see how they work. Then ask students to work from this model when designing new parachutes.

**To make it harder:** Provide a heavier load, or provide a fragile load such as an unpeeled hard-boiled egg. Increase the minimum required time for the parachute to drop. Or place cards showing different shapes (such as triangles, rectangles, and octagons) into a bag. Have each group pull out a shape card and require them to make the parachute's canopy in that shape.

# Parachute Materials Price List

Plastic shopping bags .........................................$1.00 each

Plain paper .....................................................40¢ per sheet

Facial tissues .................................................... 20¢ each

Paper towels ..................................................... 40¢ each

Waxed paper .................................................. 40¢ per foot

Foil .............................................................. 40¢ per foot

Paper bags .................................................... 20¢ each

Craft sticks .................................................... 40¢ each

Straws .......................................................... 20¢ each

Coffee filters ................................................. 80¢ each

Masking tape ................................................ 40¢ per foot

Clear tape ..................................................... 60¢ per foot

String ........................................................... 40¢ per foot

Names  _____   _____

_____   _____

# Parachute Observations

Measure and record the drop time for the control load and for all parachute drops.

| Group | Control load | Group 1 | Group 2 | Group 3 | Group 4 | Group 5 |
|---|---|---|---|---|---|---|
| Seconds to Fall | | | | | | |

1. Did our parachute take longer to fall than the control did? _____

2. Did our parachute hold together when it landed? _____

3. Was the load still attached to our parachute after it landed? _____

4. Which parachutes were the most successful:

   the ones that fell quickly or the ones that fell slowly? _____

5. Parachutes that were the most successful were similar because they all

   _____

   _____ .

6. One idea for making our group's parachute better is

   _____

   _____ .

Names _____    _____

_____    _____

# Parachute Test Results

Color each bar to show the time it took for a group's parachute to fall.

**Parachute Drop Times**

| Seconds | Control Load | Group 1 | Group 2 | Group 3 | Group 4 | Group 5 |
|---------|--------------|---------|---------|---------|---------|---------|
|         |              |         |         |         |         |         |

# Bridges

## Curriculum Connection

Science: net forces, Newton's third law of motion

## Criteria for Product

Bridges will be scored on:

- Height—the greater the clearance for a boat sailing under it, the higher the score
- Driving safety—the farther a test car travels without falling off, the higher the score
- Load rating—the greater the load the bridge can support (at least 10 weights), the higher the score

## Constraints for Challenge

- Must span a distance of 15 inches to qualify for testing
- Must use only the materials provided
- Must stay within a budget of $5.00
- Must complete each stage in the time allowed

### Challenge

Bridges connect trade routes and encourage the flow of ideas—just two of the many ways that bridges affect human societies. In this Challenge, students design, build, and test bridges.

## Materials

### Bridge Materials (for the class)

- 200 straws
- 200 craft sticks
- 200 index cards, 5 x 8 in.
- 100+ rubber bands
- 100 paper clips, any size
- 1 roll of string
- A few rolls of clear tape
- A few rolls of masking tape
- Plain paper

### Tools

- Scissors

### Test Kit Materials (for each group)

- 6 rectangular wooden blocks (1½ × 3 × 6-in. unit blocks work well)
- 1 sheet of chart paper with 1-in. rules or grids
- 1 piece of sturdy cardboard, at least 8 × 15 in. (to use as a base)
- 1 ruler or tape measure
- 1 toy boat, about 5–6 in. wide and 4 in. tall
- 1 toy car

- 60 weights in a paper cup (bolts work well)
- paper shopping bag (to hold materials kit)

### Additional Materials

- Markers
- Plain paper
- Pencils
- Rulers
- Computer with Internet access (optional)
- Projector for computer (optional)
- Pretend money, $5.00 per group

- Envelopes to hold pretend money
- Box for the "cash register"
- Playground or beach ball, underinflated
- Digital cameras, 1 per group (optional)
- Document camera (optional)
- Digital timer

**Note:** *Unless otherwise noted, provide an assortment of each item in quantities that are readily available. Make substitutions as needed.*

# Before You Begin

- Locate images of different styles of bridges. An Internet image search using the terms "arch bridge," "suspension bridge," and "beam bridge" returns good options. Bookmark, download, or print several examples for use in class discussions.
- For the test kits, look for toy boats that are close to the height of the bridge abutments (three blocks stacked up). If you find shorter boats, consider using two blocks for the abutments, instead. The idea is for the bridge clearance to be a genuine challenge.
- Use toy cars with wheels that turn easily, for the test kits.
- Assemble a test kit for each group, following the test kit materials list.
- Make a set of Job Cards (page 13), a copy of the Price List and Test Directions (page 122), and a copy of the Test Results sheet (page 123) for each group.
- For each group, place $5.00 of pretend money into an envelope.
- Make copies of the Challenge Reflection sheet (page 14) for each student.
- Place bridge materials on a table. Make signs for the price of each material, if desired.
- Set up the digital timer where students can check on it.
- Display the Criteria and Constraints where students can see them.

# 5-Step Process

**1** *Investigate*
*Session 1*
*20 minutes*

- Gather students around a desk or a table. Show students a slightly underinflated ball. Place the ball on the table with your hand on the top. Ask students:

    *Which force, push or pull, will I place on the ball if I press my hand down?*

    *What will happen to the ball?*

- Students should respond that the ball will squish, or flatten, when you push on it. Ask students why it will squish instead of moving toward the floor. Students should respond that the ball cannot go through the table. Ask students:

    *If the ball cannot go through the table, then is the table exerting any force on the ball?*

- Give students a minute to discuss this idea, and then ask students to observe what happens when you push down on the ball. Guide students to notice that the ball is not being pushed through the table. This is because the table is pushing up with a force equal to the force of your hand pushing down. The ball doesn't move down because the forces are equal. There is, however, no equal force to prevent the ball from squishing out sideways, so it does.

- Next, set a block on the table. Ask students why the block doesn't fall through the table. Guide them to recognize that even though the block is not moving or changing, there are forces acting on it—the force of gravity is pulling down and the force of the table is pushing up.
- Tell students that their next engineering Challenge will be building a bridge. Bridge designers must understand equal forces. If a bridge cannot push up with a force equal to the force pushing down on it, then the bridge will collapse. Ask students:

  *What forces are acting on a bridge?*

  *How could a bridge be designed to counteract the force of gravity pulling it down?*

  *How does something exert a force on the bridge when it walks or drives across?*

  *Other than being strong, what else makes a bridge successful?*

  *Why might a bridge need to be a certain height?*

**1** *Investigate*

*Session 2*

*20 minutes*

- Display pictures of bridges for students to observe and discuss. Include images of suspension bridges, arch bridges, and beam bridges. Invite students to comment on the similarities and differences among the bridges.
- Guide students to notice that most bridges are held up by tall columns. Tell students these columns are called abutments. Abutments are necessary to hold up the bridge. But they must be placed so they're not in the way of vehicles traveling over the bridge or under it.
- Show students how to set up the abutments that will support the bridges they will build. A diagram is shown below. Lay out a sheet of gridded chart paper. Stack three blocks on top of each other near one edge of the chart paper, such that one short edge of the blocks lines up with a rule on the paper. Count 15 one-inch spaces from that edge of the blocks, and then stack the other three blocks.

**Abutments, side view**

15 inches

- Use a tape measure to demonstrate that the distance between the two abutments is 15 inches. Explain to students that their bridges must span, or cross, this distance and be supported by the abutments. Trace around the blocks in pencil so it's easy to know where to place the abutments on the chart paper the next time.
- Point out the materials that will be available for building bridges and what their prices are. Allow students to touch and observe the materials so they have hands-on information to use in their decision making. Ask students:

  *How could you use each of the materials on the table to build a bridge?*

*Are any of these materials long enough to span the distance? If not, can you think of a way to make them span it?*

*What are some questions you have about this Challenge?*

- Review with students the Criteria and Constraints. Show them an example of a toy boat that must sail under the bridge and a toy car that must travel across it after being pushed. Then review the time limits for Brainstorm, Plan, and Build.

## 2 Brainstorm
### 30 minutes

- Tell students that they'll be thinking up as many ideas as they can for the design of a bridge that meets all the Criteria and Constraints.

- Remind students that Brainstorm is a time to let the ideas flow. They need to get all of their ideas down on paper without stopping to think about which ones they like best. Remind them to label their drawings to show which materials they plan to use. They should not consider the price of materials until instructed to do so.

- Give each student a sheet of plain paper. Allow 15 minutes for brainstorming.

- Peek over students' shoulders to check their progress as they work. Guide students to use their science knowledge as they design. Suggested questions:

    *How would these bridges stay up?*

    *Will this bridge be level enough for cars to drive across?*

    *What will keep cars from driving off the bridge?*

    *How do you know the bridge will be tall enough for boats to sail under?*

- After 15 minutes have passed, give each student a second sheet of plain paper. Tell them they have 10 minutes to decide which idea is their favorite and to draw and label it again neatly. At this point, they should check that the material prices for their favorite design are within the $5.00 budget. Let students know that it's okay to make minor changes to their design if they like, at this point.

## 3 Plan
### 30 minutes

- Ask students to list each of the four group jobs (page 12) and describe the responsibilities of each. Invite students to decide on which job they would prefer to do but tell them to have a second choice in mind, too. Let each student take the Job Card that represents the job he wishes to have, and to negotiate with others if the job he wants is not available.

- Tell students to form groups so that each group has one person with his first- or second-choice job. Each group must have someone willing to do each of the four jobs, and each group must have no more than four people. Assist students as necessary.

- Once students are in groups, set the timer for 20 minutes. Instruct them to take turns sharing their favorite ideas from Brainstorm. Remind them to consider the pros and cons of each design in order to come up with a plan that they all agree to. The plan may be one of the four favorites or a new plan based on the Brainstorm ideas.

- Walk around the room, listening in and prompting students to expand their thinking. Ask questions that guide students to make decisions based on what they have learned about forces. Sample questions:

  *Where are the forces on this bridge? Which direction is gravity pulling? Which direction is the bridge pushing against gravity?*

  *Have you included a way to keep the car from going off the side of the bridge?*

  *What do you know about equal and opposite forces that has helped you with this plan?*

  *How much weight does the bridge need to support? What can you do to build a bridge strong enough to support that weight?*

  *How do you know you have enough materials to reach across the 15-inch span? Have you measured and multiplied?*

- When groups have agreed to a plan, have each group member draw and label her own diagram of the bridge. Each student should list the materials needed and add up the total cost, showing her work.

### 4 Build

**60 minutes**

- Let the students know it's time to build the bridges. Set the timer for 60 minutes. Remind students that everyone should perform their collaborative group jobs and help build their group's bridge.

- Give each Materials Manager a test kit as well as an envelope with $5.00, and let him purchase his group's materials.

- Remind the groups to set up their abutments properly and to not handle the other test-kit materials until instructed. If available, provide digital cameras to the Recorders and direct them to take pictures of each stage of the bridge construction.

- Spend some time with each group. Ask students to tell you what their building procedure is, and to explain what they will do if it turns out their plan isn't working. Ask guiding questions to encourage cooperative behaviors and empower students to make decisions. Sample questions:

  *How is everyone involved in building the bridge?*

  *Are you working on different parts and then coming together, or are you all working on the same part at the same time? (Either approach is fine.)*

  *Will you test as you build? How do you plan to do that?*

  *What was your thinking when you designed this part?*

  *Have you had any problems so far? How did you solve them?*

- When bridges are complete, have each Recorder take a picture of another group with its finished bridge.

- Have each group slide its bridge onto the piece of sturdy cardboard for storage, before disassembling the bridge abutments. Store the bridges, on their cardboard bases, in a safe place until it's time for Test & Present.

**5** *Test & Present*

*15 minutes per group*

- Tell students that all groups will test their bridges at the same time. Give each Speaker a copy of the Test Directions. Give each Recorder a Test Results sheet. If desired, hand each Timekeeper a digital camera to take pictures during each test. Explain that the Materials Manager conducts the tests, and the Recorder writes down the results.

- Instruct the groups to set up abutments and bridges for testing. Have each Speaker read the Test Directions aloud to the group before the Materials Manager conducts each test.

- When all groups have finished testing, gather students back together for presentations. If students took photos of the tests, allow time for students to upload photos and prepare them for presenting.

- For each group in turn, help the Recorder set up the group's Test Results sheet under the document camera.

- Ask the Speaker to discuss the results of her group's test. Then have the Timekeeper display pictures of the bridge for the Speaker to talk about.

- Invite students to ask questions about the group's bridge design or their building process.

- After all groups have presented, give each student the Challenge Reflection sheet to complete.

## Opportunities for Differentiation

**To make it simpler:** Remove the budget constraint or decrease the distance of the span. Alternatively, a model bridge made of index cards can be provided to students; their task is to strengthen the bridge so it supports a load of at least 10 weights.

**To make it harder:** Decrease the budget or increase the distance of the span. Increase the height of the boat that must fit under the bridge. Require the boat to pass under the bridge while the bridge is holding a load in the middle of the span.

# Bridge Materials Price List

Straws ........................................................................ 9¢ each

Craft sticks .............................................................. 8¢ each

Index cards .............................................................11¢ each

Paper clips .............................................................. 4¢ each

Rubber bands........................................................ 18¢ for 6

String ...................................................................... 6¢ per foot

Clear tape ..............................................................10¢ per foot

Masking tape ........................................................15¢ per foot

Plain paper ............................................................7¢ per sheet

# Bridge Test Directions

1. Test the bridge height. Place the boat in the "water" and sail it under the bridge.

2. Test how safe the bridge is for cars. Place the toy car on one end and give it a push so that it travels across the bridge. Record what happens. Test the car three times.

3. Test the load that the bridge can hold. Place a paper cup in the center of the bridge. Place 10 weights in the cup and wait a moment to see if the bridge stays up. Add and count more weights, one at a time. Keep adding weights until the bridge fails. The bridge fails when the weights fall off the bridge or the bridge sinks below the top of the abutments (blocks). Note the number of weights the bridge held in the trial before it failed.

Names  _____  _____

_____  _____

# Bridge Test Results

## Criterion                                                    Score

### 1. Bridge Height                                            _____
Boat fits easily under bridge, score = 3
Boat fits tightly, score = 2
Boat touches bridge, score = 1
Boat doesn't fit under bridge, score = 0

### 2. Car Safety                                              _____
Made it across 3 times, score = 3
Made it across 2 times, score = 2
Made it across 1 time, score = 1
Did not make it across, score = 0

### 3. Bridge Load Rating                                      _____
>50 weights = 5
40–49 weights = 4
30–39 weights = 3
20–29 weights = 2
10–19 weights = 1
<10 weights = 0

## Total Score                                                 _____

# Land-Reuse Models

## Curriculum Connections

Science: human impact on the environment
Mathematics: average (mean)

## Criteria for Product

- Proposal must include a diagram, a written plan, and a realistic three-dimensional model
- Group must make a 2-minute presentation of their proposal to the community representatives
- Criteria for scoring to be determined by students during Investigate

## Constraints for Challenge

- Must use only the materials provided in the kit, but may barter with other groups to get more of the same materials
- Must complete each stage in the time allowed

### Challenge

One way to care for our Earth is to find a way to reuse a small, vacant piece of it. Students create proposals and design and build models for the reuse of a vacant lot.

## Materials

### Land-Reuse Model Materials (for each group)

- 2 sq. ft. of artificial turf, or any product that mimics green space
- 10 pipe cleaners
- 2 balls of modeling clay or salt dough, 3 in. diameter
- 1–2 c. of gravel
- 1–2 c. of sand
- 10 craft sticks
- 20 toothpicks
- 12 coffee stirrers
- 10 straws
- 3 squares of tissue paper, about 3 × 3 in., assorted colors

- 1 ft. of waxed paper
- 2 paper plates
- 10 paper clips
- 18 in. of yarn
- Slip of paper with "18 in. of clear tape" written on it
- 1 shopping bag (to hold materials kit)

### Tools

- Rulers
- Scissors
- Measuring cups

### Additional Materials

- Plastic tubs (approximately 6 × 18 × 19 in.; 1 for each group)

- Any kind of soil, enough to fill tubs to about 3 in. deep
- 1 large bag of fish-tank gravel
- Sand, about the same amount as gravel
- About 1 doz. resealable sandwich bags
- Chart paper
- Markers
- Pencils
- Plain paper
- Bag, jar, or other container (to hold slips of paper)
- Digital timer

# Before You Begin

- Locate two books about littering, recycling, or trash: one fiction and one nonfiction. Two suggestions are: *Just a Dream* by Chris Van Allsburg (fiction) and *Earth-Friendly Waste Management* by Charlotte Wilcox (nonfiction).
- Invite several parents, community members, or school staff to visit the classroom during Session 2 of Test & Present to evaluate and choose land-use designs.
- Figure out how many groups of four are in your class, and groups of three if needed. Prepare slips of paper with numbers 1, 2, 3, and so forth up to the number of groups you will have. There should be three or four slips with each number for a total of one numbered slip for each student. Place the numbers in a bag for a random drawing.
- Fill the bottom of each plastic tub with soil to about 3 inches deep.
- Scoop 1 to 2 cups of gravel into a resealable plastic bag for each group and do the same with sand. Be sure that all groups are given the same amounts of gravel and sand.
- Write "18 in. of clear tape" on enough slips of paper so that each group has one.
- Assemble a materials kit for each group, following the Land-Reuse Model materials list.
- For each group, make a set of Job Cards (page 13), a copy of the Land Reuse Plan Proposal sheet (page 131), and four copies of the Land-Reuse Model Score sheet (page 132).
- Make a copy of the Challenge Reflection sheet (page 14) for each student.
- Set up the digital timer where students can check on it.
- Display the Criteria and Constraints where students can see them.

# 5-Step Process

**1** ▸ *Investigate*
**Session 1**
*45 minutes*

- Read aloud both the fiction and nonfiction stories about taking care of the Earth.
- Invite students to think of a graphic organizer that will help to compare and contrast the two stories. Assist students in suggesting ways to complete the graphic organizer.
- As a class, discuss the themes of the books. Ask students:

  *What does it mean to be environmentally responsible?*

  *What are some ways you can help take care of the Earth?*

  *What are some ways you can work together to make the Earth and your community a better place to live?*

**1** ▸ *Investigate*
**Session 2**
*60 minutes*

- Tell students that one way to care for the Earth is to make vacant, abandoned land useful again. Inform students that in this Challenge, they will act as teams of landscape architects. The community that owns this imaginary vacant land wants to hire a company to transform it into something useful and enjoyable, but they don't know what to do.

- The community is asking all local landscape architecture companies to apply for the job of renovating the vacant land. Each company will apply by creating a proposal for what to do with the land. Each proposal must include a diagram, a written plan, a realistic three-dimensional model, and a live presentation to the community. There are no other criteria available; the community representatives have not said anything about what they have in mind for the land. The companies have to come up with their own criteria based on what they think the community would like.

- A day has been scheduled when all companies will present their plans and models to community members, who will then choose their favorites. The community spokespeople say that they will know which plan they want when they see it.

- Show students one plastic tub of soil and tell them that this is a model of the vacant land. Explain that very often when land is left vacant, it fills with weeds, trash, and other unattractive and even dangerous materials. When making their models, groups can assume that the cleanup committee has already come in and cleared the land.

- Although the project does not yet have specific criteria, it already has constraints. The community had no budget for materials to fix up the vacant land, so they gathered donations. The only materials that can be included in the plans and models are those that were donated.

- Tell students that it's safe to assume that the community has some criteria in mind for the land, even though they haven't said what they are. For example, they probably aren't looking for an ugly, poorly maintained garbage dump—or an ugly, poorly maintained anything else. So, what are they looking for? What general criteria must all designs meet, no matter how the land is used?

- Invite students to share words to describe the general criteria, and write the words where everyone can see them. Guide the discussion so that students think to include ways to measure how useful the land is, how enjoyable it is, its beauty or attractiveness, how easy the land is to maintain or keep up, and how realistic the model is.

- After all ideas are listed, assist the class with narrowing them down to the most important criteria. Help students make the criteria reasonable and measurable. For example, a realistic model has working parts, such as swings that actually swing. Although they are not being asked to build the model to scale, a park bench shouldn't be taller than a playground slide.

- Remind students that the community can use only the donated materials to renovate the vacant land. Show students the materials in one kit, which represent the donated materials. Each group must use only the materials in the kit to build their model, but if they need more of a particular material, they may arrange a trade with another group.

- Review with students the time limits for Brainstorm, Plan, and Build.

**2 ▶ Brainstorm**

*35 minutes*

- Invite students to draw a number out of the bag and then find the other students with that number to form their group. Give each group a set of Job Cards. If necessary, review job responsibilities (page 12). Instruct students to talk within their groups and decide who will take each job.

- Give each group a sheet of chart paper and markers. Ask them to consider the following questions:

    *How could this land be used?*

    *What could be built on this land that would be attractive, enjoyable, and useful to the community?*

- Remind students that brainstorming is not a time to judge ideas, but, instead, is a time to write everything down and then come up with more ideas. Later, they will go over their ideas and decide which ones are better than others.

- Set the timer for 20 minutes. Observe students as they work. Note good collaborative efforts and creativity. As needed, ask questions to guide them to ideas such as a dog park, playground, flower and butterfly garden, vegetable garden, golf course, or sports field.

- For groups that are brainstorming well, ask questions to assist them in pushing their ideas further. Suggested questions:

    *Whom do you imagine using the place you build (children, adults, families, seniors)? What do these people like to do outdoors? What could you build to help them do those activities?*

    *What kinds of animals use outdoor spaces with people?*

    *Can you attract wildlife to this space? Would that make the land enjoyable to some people? What kinds of wild animals would you want to attract?*

    *How can you make the land easy to keep up?*

    *Have you shared your ideas and listened to those of others?*

    *Have you piggybacked any of your ideas on someone else's? How did that work out?*

**3 ▶ Plan**

*30 minutes*

- Tell students they need to review their ideas from Brainstorm and come up with a workable plan that they can use to build a model. Give each group a fresh sheet of chart paper and set the timer for 30 minutes.

- Remind groups to include the following items in their plans: a color-coded sketch, labels that describe materials, and a key explaining which color stands for each material. Point out that this color-coded sketch is for the group's use. They will need to make a fresh, new diagram for the presentation, after they've completed their models.

- Let groups know that as they plan, they need to think of any items or objects to include in the design. For example, if students are brainstorming a dog park, they should list items such as fences, trees, grass, sidewalk pavers, trash cans, benches, water fountains for dogs, and leash runs.

- While checking on the plans, ask questions to clarify the students' thinking. Sample questions:

    *Is this one of your brainstorm ideas or did you mix ideas together or did you come up with something entirely new?*

    *How did you decide where to place each object in your design?*

    *How will you build each of those pieces in the model? What materials will you use?*

    *Why is that material the best for this job?*

    *Are there any other materials in the kit that would work, if you can't trade for more of that one?*

    *Why have you included that item in your design? What benefit does it offer?*

    *How do you know you can fit all these items in this space?*

    *How does your design meet the product criteria the class decided on?*

- Allow Speakers to check in with other groups to plan trades if needed. Remind them that it's better to plan ahead for a trade than to start building and then find out that no group can trade the materials you need. The Materials Managers will make the trade during Build.

**4** **Build**

*50 minutes*

- Let the students know it's almost time to create their proposals, including building the models. First, give each Recorder a Land-Reuse Plan Proposal sheet. Explain to all groups that the top part of this sheet is where they should write their plan for the presentation. It has places for describing what will be done during each phase of the proposed project. The bottom part has a table for totaling scores that will be given during the first session of Test & Present. You'll explain more about that process a little later.

- Instruct each Recorder to complete the top part of the sheet by the end of Build. When the top is complete, each member of the group reads, signs, and dates one of the name lines to show that he agrees with the descriptions that the Recorder wrote. Point out that they do not need to prepare their presentations yet. That will happen in Test & Present, Session 1.

- Have each group send their Materials Manager to pick up the materials kit and make any trades that the Speaker arranged. Tell groups that you will give them their 18-inch strip of tape when they ask for it. (Part of the challenge is the limited amount of tape, and it's impractical to put a strip of tape in the kit.)

- Set the timer for 40 minutes and tell groups to begin. Walk around the room and check in with groups. Be sure that each student is involved in some aspect of preparing the proposal (writing the plan, drawing the diagram, or building the model) and performing their group jobs as well.

- Talk with the students and ask guiding questions to focus on the details of the design. Sample questions:

    *How is each person helping?*

    *What techniques are you using to fasten the parts together?*

*Did you remember to write the plan and draw a fresh, new diagram for the presentation?*

*Did you need to trade for materials? If so, what did you need and what did you trade for it?*

*What is the most challenging part of building the model?*

- Store finished models in a safe place until it is time for Test & Present.

**5 Test & Present**

*Session 1*

**35 minutes**

- Session 1 is completed by members of the class. Confirm that the outside visitors are able to attend Session 2.

- Give each group five copies of the Land-Reuse Model Score sheet. Have them complete the top of each sheet with their names and a short description of the Proposed Use (for example, Dog Park, Playground, Ball Field). Have the Recorder write one Criterion to Score at the top of each sheet: Useful to community, Enjoyable for community, Attractive to look at, Easy to maintain, or Realistic model. (These same criteria are listed in the chart on the Land-Reuse Plan Proposal sheet.)

- Place the groups' models around the room. Behind or next to each model, put the Land-Reuse Plan Proposal and the drawn diagram. In front of each model, place all five score sheets for that group.

- Explain to students that it is time to give each other professional feedback about the designs. Everyone will visit and score each model and plan for each of the five criteria. For each criterion, they'll give a score on a scale of 1 to 5, with 5 being highest. They'll write each score in one of the boxes on the score sheet. After scoring all five criteria for one group's plan, students move on to the next. There is no particular order to follow when reviewing and scoring the plans.

- Give students 15 minutes to review and score each group's work, including their own.

- After all scores have been recorded, have each group gather up their score sheets and Land-Reuse Plan Proposal. Have students work together to calculate their average score for each criterion, and to complete the chart at the bottom of the proposal sheet.

- Tell students that when engineers write a proposal and bid for a job, they present their ideas, plans, and models to the decision makers. To give a good, solid presentation, the group needs to plan ahead. Who will speak, what will each person say, and in what order will they speak? What information will the decision makers be interested in and what can you leave out? What words can you use to persuade the community to choose your proposal? Like a commercial, the presentation needs to "sell" the group's ideas. But the presentation should not sound like a commercial. In a commercial, the actors are usually funny and informal. For this presentation, they should use their best vocabulary, speak respectfully, and have a formal tone. Also, their presentation needs to be brief. Each group has only 2 minutes to make their case that their plan is the best.

- Tell students to think about their plans' strengths. What should they emphasize about their plan that makes it better than other plans? What makes it good? How do they know it's good? Well, they have scores from other landscape architects to back up this claim. For example, if a plan received an average score of 4.6 on Attractive to look at, then the group should emphasize the appearance of their plan and model in the presentation.

- Invite students to ask any questions they have about the presentations. When all questions have been answered, give students 15 minutes to plan and practice a 2-minute presentation of their diagram, written plan, and model.

**5** ▸ *Test & Present Session 2*

*5 minutes per group, then 10 minutes for judges*

- As the judges arrive, fill them in on the criteria, the Session 1 scoring, and the procedure for presentations. Be sure they understand that after all the presentations, they will each choose one plan and explain the reasons for choosing it. Their decisions should be based on the group's presentation and model and should be independent of the other judges' decisions.

- Introduce the judges to the class. Seat the judges where they can easily see the group presentations. Explain that the groups will present their plans and models to the judges. The judges will listen to each presentation and view the models and then ask questions or make comments.

- Call groups up, one at a time, to give their presentation. Allow 2 minutes for each presentation and another 2 minutes for questions and comments from the judges.

- After all groups have presented, invite each judge to name one favorite design and to explain why it is his favorite. Explain to the class that the judges do not have to agree on one plan, although they would have to in real life, because there is only one piece of land.

- After the judges have been thanked and have left, hand out Challenge Reflection sheets and tell students to complete them.

## Opportunities for Differentiation

**To make it simpler:** Choose one option for the land-reuse plan and allow students to make plans and models for that specific task. For example, all groups design a playground, but the playground plans and models vary from group to group.

**To make it harder:** Require students to build the model to scale. Tell students that 1 inch on the model equals 10 feet on the plot of land so, for example, the containers specified in the materials list would represent a rectangular plot 180 feet by 190 feet (about 0.8 acre). Students must do research about the common sizes of things, such as baseball fields, golf courses, playground equipment, and park benches, and build these items to scale in the models.

Names _____     _____

_____     _____

# Land-Reuse Plan Proposal

**Phase 1** (Earth moving, planting trees and large shrubs, installing any hard surfaces)

_____

_____

**Phase 2** (Building enclosures and large structures)

_____

_____

**Phase 3** (Installing equipment, seating, and small structures)

_____

_____

| Criterion | Total Score | Number of Scores | Average Score (total score ÷ number of scores) |
|---|---|---|---|
| Useful to community | | | |
| Enjoyable for community | | | |
| Attractive to look at | | | |
| Easy to maintain | | | |
| Realistic model | | | |

Names _____ _____

_____ _____

# Land-Reuse Model Score

**Proposed Use** _____

**Criterion to Score** _____

Score each criterion on a separate sheet.

Score this plan or its model from 1 to 5 for the criterion listed above; 1 is low, 5 is high.

Each visitor writes one score in one box for the criterion listed above.

| | | | |
|---|---|---|---|
| | | | |
| | | | |
| | | | |
| | | | |
| | | | |
| | | | |

For group use only: Total Score = _____

# Mining Tools

## Curriculum Connections

Science: natural resources, earth materials
Social Studies: California & Alaska gold rushes, mining

### Challenge
Our earth contains minerals that we extract to make the things we design, build, and use. Students design and build mining tools and use them to mine chocolate chips from cookies.

## Criteria for Product

Guide students to include the following on the list they come up with during Investigate:

- Removes minerals quickly and easily
- Works as an all-in-one tool that can be used to dig, to remove the mineral, and to replace the dug earth
- Costs less to build than the income the group gets from the mined mineral (this is the only criterion formally evaluated for this Challenge)

## Constraints for Challenge

- May use only materials provided
- Must stay within a budget of $5.00
- May not touch the claim directly with hands
- May mine the claim for only 5 minutes during test
- Must cause as little damage as possible to the surrounding area
- Must complete each stage in the time allowed

## Materials

### Mining Tool Materials (for the class)

- 50 round toothpicks
- 50 flat toothpicks
- 50 paper clips, any size
- 25 plastic spoons
- 25 plastic forks
- 50 craft sticks
- 25 paintbrushes
- 2 rolls of masking tape
- 1 or 2 rolls of string or yarn
- 50 or more twist ties

### Tools

- Scissors
- White glue
- Stiff icing (optional)

### Additional Materials

- 2 boxes of hard chocolate chip cookies (for "claims")
- Pieces of sturdy cardboard, 4 × 4 in., to attach cookies to
- Chart paper
- Computer with Internet access
- Computer projector (optional)
- Document camera (optional)
- Pretend money, $5.00 for each group
- Envelopes to hold pretend money
- Box for the "cash register"
- Pencils
- Plain paper
- Digital timer

# Before You Begin

- Request donations of chocolate chip cookies or plan to go shopping.
- Make price tags for the mining tool materials based on the Mining Tool Materials Price List (page 139).
- Locate images of mines and mining tools. Images may be historical or modern, but the collection should include some simple, hand-operated tools. For an Internet image search, combine the word "mine" or "mining" with the name of the mineral being mined, such as "gold mine" or "coal mining." Add the word "tool" to locate images of tools. Bookmark or download images for use during Investigate.
- Locate and print a black and white outline map of the United States or of your home state. An online search, especially of teacher-oriented websites, is useful.
- Make the cookie mines stable and secure by taping or gluing claims to small pieces of corrugated cardboard. If white glue does not work, try stiff icing, such as the icing used to hold together gingerbread houses.
- Set up an envelope with $5.00 of pretend money for each group.
- Make a set of Job Cards (page 13) and a copy of the Mining Tool Income Statement (page 140) for each group.
- Make a copy of the Challenge Reflection sheet (page 14) for each student.
- Set up the digital timer where students can check on it.
- Display the Criteria and Constraints where students can see them.

# 5-Step Process

**1** **Investigate**
*Session 1*

*45 minutes*

- Show students one of the pictures of a mine that you found. Ask students:

  *What does this image show?*

  *What do you know about mining?*

  *Where do prospectors look for minerals? How do they know where to look?*

  *How do you think miners get at the minerals if they are in the earth?*

  *What tools do miners use to get minerals out of the earth?*

- Let students know that today, in modern times, there are many advanced tools to help locate mineral resources. But historically, most miners did not know where to look unless a sample of the mineral was found above ground. Even then, they could not be sure how much of the mineral was in the earth in that place.

- Display the outline map of the United States or of your state. Explain to students that, when a mineral is found, it is also important to know who has the right to mine it. When prospectors or companies mine for minerals, they purchase or lease one or more claims. They are allowed to dig only on their own claims, and no one else is allowed to dig on their claims. Draw lines to divide a state into smaller sections to demonstrate how a map of claims might look.

- Tell students that drawing the outline on a map is one way to model a mining claim. Another way is to use an object that you can touch and work with.

- Hold up a chocolate chip cookie and explain that for their next Challenge, these cookies will be models of their mining claims. The cookie part represents dirt and rocks while the chocolate chips stand for the mineral they'll be mining.

- Pass out cookies so that pairs or small groups of students each have a cookie to inspect. Ask students:

  *How is this cookie like a real mine and how is it different?*

  *Is it possible to remove the chocolate chip "mineral" from this mine with your bare hands? How well does that work?*

  *Are all of the mineral pieces at the surface of the mine or are some buried?*

  *How can you locate the buried mineral pieces?*

  *What might help get the mineral out?*

  *Do you think it's okay if the earth in the mine breaks into several pieces?*

  *Where should the leftover pieces of earth go when you're finished mining?*

- Display the images of common handheld mining tools that you located during Before You Begin. Explain that these types of tools have been used in combination for years, with different tools for different stages of mining. Tell students that for their next Challenge, they will design, build, and use a new all-in-one mining tool to get at the mineral (chocolate chips) and remove it from the earth (cookie part) while disrupting the earth as little as possible.

**①** *Investigate*
*Session 2*
*45 minutes*

- Set the Mining Tool materials out on a table along with their price tags.
- Remind students that their Challenge is to build an all-in-one mining tool. An all-in-one tool digs into the earth, removes the mineral, and replaces the earth that was moved.
- Tell students that for this Challenge, they will be given the constraints, but they will come up with the criteria based on discussions from the previous session. Point out that although they have a budget for materials to build the tools, they will also earn money from selling the minerals they mine.
- First read aloud the constraints and discuss them so students know what they are. Explain that the cookie claims will be glued down to help keep them in place. Miners (students) will only be allowed to touch the claim with the tool they build—not with their hands. Also, they must do as little damage as possible to the surrounding earth while they dig for the mineral. This includes keeping all earth material on their claims and putting back any materials that are moved.

- Next, guide a discussion with students about the criteria they'd like to set for this Challenge. You could guide the discussion toward criteria similar to the ones listed on page 133. Suggested questions for the discussion:

  *Which would be better, a tool that takes a long time to use or a tool that makes the work go by more quickly?*

  *Is quicker better if there's more cleanup to do afterwards?*

  *Does it matter how much it costs to build the tool? What if the tool costs more than you can earn from selling the mineral?*

- After students have come up with the criteria, write them down and post them where everyone can see them.

- Show students the materials they are allowed to purchase for building their all-in-one mining tools. Ask students:

  *What would be a good material for digging, without disturbing too much earth?*

  *Can any of these materials be used in more than one way?*

  *Why might you choose a rounded toothpick over a flat toothpick, or vice-versa?*

  *How could you use a spoon?*

  *What are different ways you might combine the materials into an all-in-one tool?*

  *What is the paintbrush good for?*

- Review with students the time limits for Brainstorm, Plan, and Build.

**2 ▶ Brainstorm**
*30 minutes*

- Invite students to choose a job that they would like to do for this Challenge: Recorder, Timekeeper, Speaker, or Materials Manager. They should have a second-choice job in mind, too.

- Once students have chosen jobs, help them form groups of four so that each group has someone to do each job. If there are any groups of three, then one student takes two jobs.

- Give each group a sheet of chart paper. Ask them to work together to think of as many different ways as possible to combine the materials that are available into an all-in-one tool for mining chocolate chip mineral pieces. Everyone in the group should offer ideas and write on the paper. Set the timer for 20 minutes.

- Walk around and listen to conversations. Remind students to be respectful of each other as they talk about the use of materials to create a mining tool. Suggested questions:

  *How many different ideas have you come up with?*

  *How will you hold all the parts of the tool together?*

  *How could you use the craft stick?*

  *What does this piece of the tool do that the other pieces do not do?*

- When the time is up, remind students to stop brainstorming.

**3** *Plan*
*30 minutes*

- Instruct students to work together to decide on an all-in-one tool design. They need to confirm that the total cost of the chosen design is within budget. Set the timer for 30 minutes.

- Check in with each group as students work. Join in on conversations. As you view the plans-in-progress, ask questions to help students make smart decisions. Sample questions:

  *Why did you decide on this design? What was your thinking?*

  *How does this plan make it easy to leave the earth close to the way you found it?*

  *Of the designs that you considered, is this one of the more expensive ones or less expensive ones?*

  *What properties does this tool have that made you choose it?*

  *Which part of this design digs? Removes the mineral? Replaces the soil?*

  *Can the tool work without the miner touching the cookie claim?*

- Tell the class that each student in the group should draw and label the plan that the whole group agreed to. If the plans do not match, then students need to discuss the differences and agree on one design

**4** *Build*
*45 minutes*

- Let the students know it's time to create the mining tools. Set the timer for 45 minutes. Hand each Materials Manager an envelope with $5.00 of pretend money. Hand each Recorder a copy of the Mining Tool Income Statement.

- Invite the Materials Managers to go shopping for the materials that the groups listed on their plans. The Recorders should fill in the cost of materials that the Materials Managers purchase. This information goes on the Income Statement. Once a group has its materials and the costs have been recorded, then the group may begin working.

- As you circulate, check that all students are helping to build the tools and also doing their group jobs.

- Visit each group and talk with the students as they build. Ask guiding questions to keep students on task. Sample questions:

  *What is each person doing to help build?*

  *How will you use the tool to remove the mineral?*

  *How will you use the tool to put everything back to the way it was, as best you can, when you're finished?*

  *What is the most challenging part of this Challenge?*

- Store all finished mining tools in a safe place until it is time to Test & Present.

**5** *Test & Present*
*10 minutes per group*

- Call on one group at a time to present. For each group, place a cookie claim, glued to cardboard, under the document camera, if you are using one. The claim should be within reach of the presenters and projected onto a large screen. If no camera is available, have other students in the other groups gather around the presenting group.

- Invite the group's Speaker to talk briefly about their mining tool. Each presentation should include information about why the group chose specific materials, how the tool works, and how each member of the group worked together as a team.
- Ask the Materials Manager to choose one other person from the group to help demonstrate and test the tool. Invite them to bring their tool up to the claim.
- Remind students that as they mine for minerals, the only thing that may touch the claim is the tool. One person may, however, hold the cardboard base in place while the other uses the tool.
- Let students know they have 5 minutes to mine as much mineral as possible. Tell the Timekeeper to set the timer for 5 minutes. Each whole mineral chip mined earns the group $5.00. If the mineral breaks, then all the broken pieces together count as one chip.
- After 5 minutes, mining stops. The Recorder adds up the total number of mineral chips mined and then multiplies to determine the total income. As the next group sets up, the Recorder completes the rest of the Mining Tool Income Statement.
- After all group presentations are completed, have the groups share their results. Discuss which mining tool was the most successful based on the Mining Tool Income Statement. Then ask students to discuss which group's result followed the Constraints most closely. Continue the discussion of which tool was most successful. Invite all students to speak up, to share a variety of opinions, and to support their ideas and opinions with the scores and facts. It is possible that the tool that made the most profit was not the most successful if it disturbed the earth too much—but the students must decide this as a group.
- Provide each student with a copy of the Challenge Reflection sheet to complete.

## Opportunities for Differentiation

**To make it simpler:** Allow students time to try items from the materials list on chocolate chip cookies before designing their all-in-one tool. Demonstrate how tools can be combined by placing items parallel to one another, perpendicular to each other, or at other angles.

**To make it harder:** During the test, count only mineral chips that are still in one piece after mining. Add a criterion that the tool must be durable, and have students do the test multiple times to check for wear and tear on the tool. Or provide a variety of mining claims—that is, cookies with different sizes and quantities of chips—and different costs, and require students to use part of their material funds to stake their claims.

# Mining Tool Materials Price List

Round toothpick ....................................................$2.00 each

Flat toothpick.........................................................$1.00 each

Paper clip ..............................................................$2.00 each

Craft stick .............................................................$1.00 each

Paintbrush.............................................................$3.00 each

Tape......................................................................$1.00 per foot

String or yarn .......................................................$0.25 per foot

Twist ties...............................................................$0.25 each

Names  _____  _____

_____  _____

# Mining Tool Income Statement

**Income**

| Number of chips mined | x $5.00 per chip (show work) | Total Income |
|---|---|---|
|  |  | $ |

**Expense**

| Material | Number of items | Cost per item* | Total cost (number of items x cost per item) |
|---|---|---|---|
|  |  |  |  |
|  |  |  |  |
|  |  |  |  |
|  |  |  |  |
|  |  |  |  |
|  |  |  |  |
| **Total Expense** |  |  | $ |

*\* Costs are listed on Mining Tool Materials Price List*

Total Income (from above) =          $_____

Total Expense (from above) =          $_____

Total Profit or Loss (income minus expense) =          $_____

# Oil-Spill Cleanups

## Curriculum Connection

Science: natural resources, environmental hazards

## Criteria for Product

Oil-spill cleanup system will be scored on:

- Cost—lower-cost systems score higher
- Containment—keeps spilled oil from spreading
- Removal—removes spilled oil from the water
- Realism—could be used in the real world
- Safety—animals can escape from it

## Constraints for Challenge

- System must fit in the oil-spill model tub
- Must use only the materials provided
- Must stay within a budget of $225.00
- Must complete each stage in the time allowed

### Challenge

Oh no! No one wants to see slick black blobs from an oil spill at a favorite beach. Students design, build, and test a system to clean up an oil spill.

## Materials

### Oil-Spill Cleanup Materials (for the class)

- Plastic bottles
- Plastic food-service gloves
- Egg cartons
- Foam cups
- Craft feathers
- Nylon knee-highs
- Several rolls of paper towels
- Foam trays
- Sponges
- Pipe cleaners
- Plastic pipettes or droppers

- Plastic spoons
- Craft sticks
- Plastic plates
- Sandwich bags
- Aluminum foil
- Plastic wrap
- Sturdy string
- Clothespins
- Masking tape
- Duct tape

**Note:** *Provide materials in quantities that are readily available.*

### Tools

- Scissors
- Hole punches
- Staplers

- Penknife (teacher use only)
- Glue gun (teacher use only)

### Additional Materials

- Chart paper
- Markers
- Pencils
- Plain paper
- Clear plastic tubs with snap-on lids, 1 for each group
- 3–4 qts. of cooking oil
- 1 lb. of cocoa powder

- Large container and spoon
- 12-oz. paper cups, 2 for each group
- Digital timer
- Pretend money, $225.00 for each group
- Envelopes
- Box for the "cash register"
- Water

# Before You Begin

- Gather books and online resources about oil-spill cleanups for students to use during Investigate, from the Internet and local libraries. Look specifically for information about cleaning oil that has spilled in the water; information about cleaning oil from animals and coastlines is not relevant to this Challenge. Suggested bibliography for Internet: Twin Cities Public Television (tpt.org), search for Newton's Apple video "Oil Spills"; Scholastic (scholastic.com), search for "Gulf Oil Spill Recovery Special Report"; National Wildlife Federation (nwf.org), "The Big Oil Spill." For books: *The Exxon Valdez Oil Spill* by Peter Benoit (a True Book); *Oil Spill!* by Melvin Berger (Let's-Read-and-Find-Out Science 2); *Oil Spill: Disaster in the Gulf* by Scholastic.

- Pour water into one tub so that it is half full.

- Combine the cooking oil and cocoa powder to create a dark, oily mixture. Fill one 12-ounce cup with the mixture for Investigate Session 1. Fill cups for each group before Test & Present.

- If desired, make price tags for the materials, using the Price List on page 148. Otherwise, make a copy of the Price List to post with the materials, in addition to a list for each group.

- For each group, prepare an envelope with $225.00 in pretend money.

- Make a copy of the Oil-Spill Cleanup Price List (page 148) and Test Results (page 149) for each group.

- Make a copy of the Challenge Reflection sheet (page 14) for each student.

- Set up the digital timer where students can check on it.

- Display the Criteria and Constraints where students can see them.

# 5-Step Process

 **Investigate**
**Session 1**
*30 minutes*

- If you can safely set up a document camera over the tub that is half-filled with water, then do so. If there is any risk from using an electrical device near water in your setup, then instead of using the camera, place the tub where students can gather around it. Tell students that this is a model, and ask for ideas about what the model might represent. When a student guesses that it is a model ocean, confirm that this is correct.

- Ask students what they know and have heard about oil spills. Allow time for students to share their knowledge and experiences, and then invite students to think of ways that an oil spill can happen. Confirm that one common way is when an oil tanker (a ship that carries oil to where people need it) gets a leak, either because it is faulty or because it runs into something that tears a hole in it. When tankers leak, oil spills out across the ocean.

- Tell students that they will observe a model of an oil spill on the ocean. It is important that they observe how the oil and water interact as they watch the model. You will write their observations on the board or on chart paper as they share them.

- Hold up the cup of oily mixture and say that this material is a model of crude oil. Explain how it was made and assure students that it is safe to handle.

- Choose one student to be the "tanker" that spills the oil on the ocean. The student may add the oil to the model using any method he chooses, as long as no oil or water falls outside of the model. Remind students that it is important that they observe what happens as they add the oil, and then let the student add the oil. Ask students questions to support their observations:

  *What is happening to the oil and water?*

  *Where is the oil going as it is poured? To the top? To the bottom?*

  *What happens when the oil touches the sides of the model?*

  *How does the oil look?*

  *Where does the oil end up once the water stops moving?*

  *What does the oil do after it's been poured?*

- As students call out observations about what they are seeing, write them on the board or on chart paper. Students should note that oil floats on top of the water. The oil continues to spread out farther and farther, even after it is poured. When small blobs of oil touch, they join together into larger blobs. Encourage students to describe anything they see happening.

- When students have stopped making new observations, lead a class discussion about what they saw.

  *Why is an oil spill a problem?*

  *What can people do to fix this problem?*

  *What do you need to know about oil and water to solve this problem?*

  *What materials might help solve this problem?*

- Place the lid on the oil-spill model and store it in a safe place until Session 2.

**Investigate**

**Session 2**

*45 minutes*

- Place all oil-spill cleanup materials and tools on a table, with price tags if available.

- Bring out the oil-spill model and remove the lid. Ask students to recall what they did, observed, and discussed during Session 1. Let them observe how the model looks different than it did yesterday; for example, the oil has spread out over the whole surface and may be clinging to the walls of the container.

- Ask students if they can guess what their next engineering Challenge will be. Confirm that the Challenge is to design an oil-spill cleanup system. A *system* is a combination of materials and procedures that work together to form a solution. Each group will test their system on an oil-spill model that they make during Test & Present.

- Inform students that before embarking on this task, they need to think about different procedures for cleaning up an oil spill. Ask students:

  *Do you want to invent something that keeps the oil in one place so it doesn't spread any farther?*

*Should you design something that removes the oil from the water?*

*Is it possible to invent something that does both?*

- Allow 20 minutes for students to work with the materials you gathered about oil-spill cleanups, either at their desks using print materials, or online using computers. The goal is for students to learn about things people have already done to try to solve this problem before making plans for their own cleanups.

- After students have completed their research, gather them together and allow 5 minutes for them to share information.

- Show students the materials available for this Challenge. Ask questions to get students thinking about how they could set up an oil-spill cleanup system, using the available materials. Suggested questions:

    *What could you use to contain the oil and keep it from spreading?*

    *What could you use to remove oil from the water?*

    *Where could you put the oil once you remove it?*

    *How will the equipment stay in place?*

    *What questions do you have about the Challenge?*

- Read the Criteria and Constraints and discuss the meaning of each item. Make sure that students recognize that the word *animals* includes not just mammals, but also birds, fish, turtles, and so forth. Let students know that they will be allowed to use only the materials on the table and that they have a budget of $225.00 to spend. Point out that $225.00 sounds like a big budget, but the prices of the materials are high for this Challenge.

- Review the time limits for Brainstorm, Plan, and Build.

## ② Brainstorm
### 30 minutes

- Tell students that they'll be thinking up as many ideas as they can for the design of a system that can contain and clean up an oil spill on the ocean surface.

- Give each student a sheet of plain paper. Remind students that Brainstorm is a time to let the ideas flow. They need to get all of their ideas down on paper without stopping to think which ones they like best. Ask them to label their drawings to show the materials they plan to use.

- Allow 15 minutes for brainstorming. Peek over students' shoulders to check their progress as they work. Guide students to use their science knowledge as they design. Suggested questions:

    *How will you know if your oil-spill cleanup is successful?*

    *What properties do these materials have that might help build an oil-spill cleanup system?*

    *How are these materials alike? How are they different?*

    *What could you use this item for? How about this one?*

    *What are you using to contain the oil?*

    *What are you using to remove the oil? Where are you putting the oil you remove?*

*Will the equipment float? Why is it helpful if it does?*

*How would the equipment be put in place in the real world?*

- After 15 minutes have passed, give each student a second sheet of plain paper. Tell students they have 10 minutes to decide which idea is their favorite and to draw and label it again neatly. Let them know that it's okay to go ahead and make minor changes to their design at this point, if they like.

### 3 ▸ Plan
**45 minutes**

- Tell students that they will form groups based on the similarity of designs.

- Give students 5 minutes to walk around the room with the sketches of their favorite ideas. They need to look at each other's sketches and briefly discuss them, so they can find people who have designs similar to their own. Once they have three or four people in the group, they should sit down together.

- Visit each group as they sit down. Hand each student in the group the job card for the job she would be least likely to choose (Timekeeper, Recorder, Speaker, or Materials Manager). Let students know that you assigned those jobs to help stretch them in new directions.

- Provide each group with a copy of the Price List, a sheet of chart paper, and markers.

- Point out the prices for Tool Rental. Explain that the 5-minute tool rental charge applies even if they use a tool for only 1 minute. They're also charged for the extra time if they forget to return a tool. So they should think about all the things they may need each tool for, how many times they'll need each tool and for how long, and then include those costs in their plans.

- Introduce yourself as the Contractor. You will be available for hire during Build. Groups can hire you to do the jobs listed, as well as any other job they cannot do themselves. Groups need to include Contractor costs in their plans.

- Tell students to use their ideas from Brainstorm to come up with one group plan. Each plan needs the following items: a description of how the system works, a sketch with labels showing materials, a materials list, and the total cost to build the system, including the cost of materials, tool rentals, and Contractor hires.

- Set the timer and give students 30 minutes to plan. Visit groups as they work. Encourage groups to draw up detailed plans. Suggested questions to ask:

  *What does each part of this system do?*

  *How do you know the system will fit in the oil-spill model?*

  *What holds everything together? What keeps it in place?*

  *What powers the system? Does a person handle it, or is there another way it operates?*

  *Did everyone have a voice in this plan? Were everyone's suggestions considered?*

### Build
**60 minutes**

- Let students know it's time to build the oil-spill cleanup systems.
- Remind students they will be testing their solutions in a tub the size of the model they saw earlier. They should keep this in mind as they build.
- Call on student volunteers to review the procedure for Build, as well as the responsibilities of each group job (page 12). Be sure students mention the procedure for revising plans: everyone agrees to the change; the Speaker gets approval from the teacher; the Recorder revises the plans.
- Set the timer for 45 minutes and point it out to the Timekeepers.
- Give each Materials Manager $225.00. Act as the Distributor as they purchase materials listed in their groups' plans. When all groups have their materials, switch roles from Distributor to Contractor.
- Check in regularly with each group between your Contractor jobs. If groups are stuck, do not offer suggestions. Rather, ask questions to guide students to work through challenges and misconceptions. Sample questions:

  *How have you divided the work so that everyone has something to do and you finish building on time?*

  *What are the steps for making your oil-spill cleanup system work? Tell me about each step.*

  *Is there any way for you to test the system before you have a model oil spill to work with?*

  *Has anything not worked the way you planned? What did you do to work through that challenge?*

- As each group finishes, give the Recorder a Test Results sheet and instruct them to complete item 1, on cost.
- Store oil-spill cleanup systems in a safe place until it's time for Test & Present.

### Test & Present
**10 minutes per group**

- For each group, half-fill a tub with water and prepare a 12-ounce cup of model crude oil. Set the models out around the room. Wait to hand out the cups of crude oil until it is time for each group to present.
- Call on groups one at a time. Have each group gather around their oil-spill model.
- Ask the Speaker to say how much the oil-spill cleanup system cost to build and to describe how it works. Then, hand the Materials Manager a cup of model crude oil to spill into the model ocean. Allow the water to stop moving, and then give the group 5 minutes to use their oil-spill cleanup system to contain and clean up the oil spill.
- After 5 minutes, remove the equipment from the water. Ask the class to determine how well the oil was contained or how much it spread, and how much of the oil has been removed, based on the appearance of the model. Have the Recorder complete the Test Results sheet.

- Allow 2 minutes per group for additional questions and comments from the teacher and students.
- When all groups have completed their presentations and the models have been cleaned up, distribute copies of the Challenge Reflection sheet for each student to complete.

## Opportunities for Differentiation

**To make it simpler:** Allow students to choose between containing the oil or removing the oil, rather than doing both. Make the math simpler by reducing the budget and prices.

**To make it harder:** Require students to explain what each part of their design represents, and what it would be made of, in the "real world." They must include a report that lists actual materials and sizes, based on research about real materials used in oil-spill cleanups.

# Oil-Spill Cleanup Price List

## Materials

Large plastic bottles ..................... $37.50 each

Medium plastic bottles ............... $25.00 each

Small plastic bottles ..................... $12.50 each

Plastic gloves ............................... $10.00 each

Egg cartons ...............................$5.00 per cup
(cut upon request)

Cups ..............................................$10.00 each

Feathers ..........................................$1.50 each

Nylon knee-highs .......................... $7.50 each

Paper towels ................................. $5.00 each

Foam trays ................................... $12.50 each

Sponges .........................................$17.50 each

Pipe cleaners ................................ $2.50 each

Spoons ........................................... $3.50 each

Droppers ........................................ $4.00 each

Craft sticks ................................... $3.00 each

Plates .............................................. $3.00 each

Sandwich bags ............................. $3.50 each

Aluminum foil ........................... $12.50 per foot

Plastic wrap ..............................$7.50 per foot

String ........................................$12.00 per yard

Clothespins ................................... $5.00 each

Masking tape .......................... $7.50 per yard

Duct tape ................................$12.00 per yard

## Tool Rental
(Minimum charge per session is 5 minutes.)

Scissors .......................... $20.00 for 5 minutes

Stapler ............................$15.00 for 5 minutes

Hole punch ....................$15.00 for 5 minutes

## Contractor Hire .................. $25.00 per job
(Ask about other jobs.)

• Cut bottles, boxes, egg cartons

• Poke holes

• Use glue gun

Names _____   _____

_____   _____

# Oil-Spill Cleanup Test Results

## Criterion                                                    Score

### 1. Cost                                                     _____
Cost $75 or less, score = 3
Cost $76–$150, score = 2
Cost $151–$225, score = 1
Cost more than $225, score = 0

### 2. Contains oil                                             _____
Oil completely contained, score = 3
Oil mostly contained, score = 2
Oil slightly contained, score = 1
Oil not contained, score = 0

### 3. Removes oil                                              _____
Removes most oil, score = 3
Removes some oil, score = 2
Removes a little oil, score = 1
Does not remove oil, score = 0

### 4. Realism                                                  _____
Definitely could be built in the real world, score = 3
Might be built in the real world, score = 2
Would be hard to build in the real world, score = 1
Could not be built in the real world, score = 0

### 5. Animal safety                                            _____
Most animals can escape from all parts, score = 3
Most animals can escape from some parts, score = 2
Some animals can escape from some parts, score = 1
Most animals could be trapped by most parts, score = 0

### Total Score                                                 _____

# New Businesses

## Curriculum Connections

Mathematics: operations
Social Studies: economics

## Criteria for Product

- Needs a catchy name
- Must be something children want or can use
- Must be priced to produce a minimum profit of 20¢ per item
- Must have at least 10 of the items available for sale at a craft fair
- Must use at least one marketing method to promote the product at the craft fair

## Constraints for Challenge

- Must use only the materials provided
- Must take no more than 5 minutes to assemble each item
- Must be made in the classroom
- Must not require packaging
- Must complete each stage in the time allowed
- All work must be done by business owners (members of the group). No outside help is allowed.
- Total craft-fair space is one table and two chairs

## Materials

### New Business Product Materials (for the class)

- Beads
- Yarn
- String
- Hair clips & pins
- Key rings
- Small foam cutouts
- Craft foam sheets
- Clear plastic cups
- Rolls of duct tape
- Buttons
- Pom-poms
- Craft sticks
- Pipe cleaners
- Erasers
- Straws
- Tissue paper
- Colored cellophane
- Cardboard tubes

### Tools

- Scissors
- Rulers
- Pencils
- Permanent markers
- Clear tape
- Glue
- Paint
- Paintbrushes

### Additional Materials

- Chart paper
- Markers
- Pencils
- Plain paper
- Index cards
- Construction paper
- Pretend money
- Envelopes
- Boxes for "cash boxes"
- 5-lb. bag of white sugar
- 72 oz. of lemon juice
- Water
- Pitchers
- Measuring cups
- Mixing spoons
- 8-oz. cups
- Straws
- Digital timer

# Before You Begin

- Locate a book about lemonade stands, such as *Olivia Opens a Lemonade Stand* by Kama Einhorn, *The Lemonade War* by Jacqueline Davies, or another book that is readily available.
- Review the New Business Product Materials list and make substitutions as desired. Request donations of materials or plan to go shopping. Most materials can be purchased at craft stores. Revise the New Business Product Materials Price List (page 159) as needed and make one copy for each group.
- Recruit at least one other adult to help fulfill material orders during Build and to help supervise the craft fair during Test & Present.
- Invite another class to participate as customers at the craft fair during Test & Present.
- Place $5.00 in pretend money in an envelope for each customer to make purchases.
- Place $20.00 in pretend money in an envelope for each group to purchase product materials and make change during the craft fair.
- Make a set of Job Cards (page 13) and a copy of the Lemonade Stand Word Problem (page 158), Product Information sheet (page 160), Materials Order Form (page 161), and Income Statement (page 162) for each group.
- Make a copy of the Challenge Reflection sheet (page 14) for each student.
- Set up the digital timer where students can check on it.
- Display the Criteria and Constraints where students can see them.

**Get Set**

# 5-Step Process

**Investigate**

**Session 1**

**60 minutes**

- Read aloud a book about lemonade stands. Depending on the length, you may read the whole book in one session or you may read excerpts from a longer book. As a class, discuss the idea of a lemonade stand business. Let students share any experiences they've had with lemonade stands or other kinds of businesses.
- Explain that the difference between what a business makes and what it spends is its profit or loss. A business makes a profit when it takes in more money than it spends. A business has a loss when it spends more money than it takes in. If your students are familiar with positive and negative numbers, describe profit as a positive value and loss as a negative value.
- Ask students:

    *What are the costs of running a lemonade stand?*

    *Does anyone get paid to make and sell the lemonade?*

    *How does a lemonade stand make a profit?*

    *How can you keep track of costs and sales to find out how much money the business makes?*

- Display the Lemonade Stand Word Problem (page 158) where students can see it. Ask students to figure out the cost for one serving of lemonade. As students work, discuss how they are solving the problem. Some students will realize that they are missing a piece of information: they need to know how long it takes to make lemonade. Ask them to think about the following:

> *How much would you charge for a serving of lemonade?*
>
> *How many servings of lemonade would you need to sell at that price in order to cover the cost of the materials to make a pitcher of lemonade (12 servings)?*
>
> *What will happen to the cost of a pitcher of lemonade when you add in the cost of labor (the workers)?*

- Provide each group with the materials listed on the Lemonade Stand Word Problem sheet, except for cups and straws. Tell students they have 1 minute to discuss within their groups how they will divide up the work needed to make the lemonade.

- Provide each group with a digital timer or stopwatch, and have each group time how long it takes them to make a pitcher of lemonade. After all groups have finished making lemonade, provide paper cups and straws so that students can enjoy the results of their efforts.

- While they enjoy their lemonade, have students figure out the cost of making the product. Define product for students as an item that is made for sale. In this case, the product is lemonade. The cost of the product is the sum of all costs that go into making it—materials (including packaging) and labor. Students may have already figured out the cost of materials for the lemonade: $1.92 for 12 servings, or $0.16 (16¢) per serving. The cup and straw bring the cost up to $0.18 (18¢) per serving.

- Tell students that they now need to figure out the cost of labor and add it to the cost of materials. Explain that *labor* is another word for *work*. Remind students that it costs 10 cents per second to pay all the workers (not 10 cents per worker per second, but 10 cents per second for all the workers together). One way to find the total cost is to first multiply the number of seconds it took to make the lemonade by 10 cents (cost of labor per second). Then, to figure out the labor cost per serving, students divide the total labor cost by the number of servings of lemonade made—in this case, 12. Then, they need to add the labor cost per serving to the materials cost per serving to find out the total cost of a serving of lemonade.

- After students have added the labor costs to the materials costs, ask:

> *Do you want to keep your lemonade at the same price or change it, after adding in the cost of labor?*
>
> *Is there any way to reduce the cost of labor? Can you change the time required, or can you change how much you make in the same time?*

- Explain to students that finding the cost of lemonade was for practice. They'll be using that knowledge in their next Challenge. Make arrangements to store or share the leftover lemonade, if desired.

## 1 Investigate
*Session 2*
*30 minutes*

- Display product materials on a table with copies of the New Business Product Materials Price List.

- Tell students that it is time to apply what they learned from making lemonade to the design and launch of a new business. The business will create a product that will appeal to children their own age, using any of the materials on the table. Then all of the businesses will display their products at a craft fair, where members of another class can buy them using pretend money. Their goal is to have a successful business that ends up with a good profit.

- Invite students to offer ideas about the different things they need to think about and do, as they start their own businesses. Write their ideas on the board. Guide students to include that they need to know what product to sell, the cost of materials, how to assemble the product, the cost of labor, a good pricing strategy, and ways to make people aware of the product and convince them to buy it (marketing and sales). Suggested questions:

  *What kinds of products would children want to buy? Toys, games, something to wear, something to collect? Something useful for school or after-school activities?*

  *Once you come up with an idea for a product, what do you need to consider before you can make it? For example, what materials should you use? How will you hold the materials together? How much will the materials cost? How will you assemble it?*

- After a good discussion, read aloud the Criteria and Constraints. Define marketing for students as something that is done to make sure people know about a product and to persuade them to buy it. Examples of marketing efforts include putting up signs, making flyers, placing advertisements, and having sales people talk to customers.

- As a class, summarize the discussion by listing a step-by-step guide to engineering a new business. Make sure students include everything they need to do in order for their businesses to be successful in this Challenge. Refer to the Criteria and Constraints again, as needed.

- Tell students that for this Challenge, labor will cost $5.00 for the 45-minute Build time, regardless of whether a group has three or four people, or how many items the group makes. The labor cost per item will be $5.00 divided by the number of items the group makes during Build.

## 2 Brainstorm
*20 minutes*

- Before starting the official Brainstorm session, allow a few minutes for open discussion as students examine the materials and check their costs.

- Explain to students that it is time to Brainstorm. Give each student a sheet of plain paper. Tell them they'll each work in whatever way is best for them. Set the timer for 15 minutes and allow students to brainstorm. As they work, observe their processes and ideas. Ask students:

  *What is your brainstorming process? Are you writing up one idea, four ideas, or lots of ideas? Why did you choose to work that way? How are you thinking up new ideas? Are you looking at the materials first for inspiration, thinking of products and then looking at the materials, or taking some other approach?*

- When the timer goes off, ask students to decide which of their product ideas they would most like to make for their new business and circle it.

### 3 Plan

*40 minutes*

- Tell students to walk around the room, discuss their designs with other students, and make groups of three or four with others who have similar product ideas. Once they have enough people in a group, they should sit down together.

- As each group sits down, give them a set of Job Cards. Assign each student in the group the job he would be least likely to choose (Timekeeper, Recorder, Speaker, Materials Manager) and hand him the appropriate card. If they ask, tell students that you assigned those jobs in order to help them stretch their skills in new directions.

- Provide each group with a sheet of chart paper and a set of markers. Set the timer for 20 minutes. Tell students that the whole group should work together to draw one plan. The plan should include the following:
  - Diagram of the finished product with all materials labeled
  - Written plan for assembling the product during Build
  - Total cost of materials to assemble at least 10 items
  - Ideas and plans for marketing the product

- As you go around the room talking with the students, ask questions to extend their thinking and push them to be innovative thinkers. Suggested questions:

  *What is your product? Why would children want it?*

  *Do you have a name for your product yet?*

  *How long do you think it will take to make one product?*

  *How many do you think you can make in 45 minutes?*

  *What will you do to market the product?*

  *Does your design meet all the Criteria and Constraints?*

- After the timer goes off, hand each Recorder a copy of both the Materials Price List and the Materials Order Form. Hand each Materials Manager an envelope of pretend money. Tell groups that they have 10 minutes to fill out their materials order form and turn it in to you with the correct payment (line D). They should leave lines E and F blank, as they do not yet have the information to complete them. You may prefer to walk all groups through completing the sheets as a class.

- Fill each group's materials order before the next stage, so it is ready for the Materials Manager to pick up at the start of Build. Place items in a shopping bag with the group's completed order form.

**Build**
*Session 1*
*60 minutes*

- Deliver each group's completed materials order with its form to the appropriate Materials Manager. Remind students to pick up any tools they may need to assemble the product. Tell groups to hold onto their Order Forms, as they'll need them again later.

- Tell students they'll have 10 minutes to make one or two samples of their product. Groups should time how long it takes to make each sample. This is the part of Build when they may make changes to their product or to their manufacturing process, if necessary. Set the timer for 10 minutes of sample assembly. Walk around the room, observing how students proceed. Suggested questions:

  *Is the product coming out the way you imagined? Do you need to make any changes to it?*

  *What was your plan for assembling the product? Did you plan on an assembly line or was each person making one item?*

  *How long did it take to make one sample? Did the second sample go faster?*

  *Will 45 minutes be enough time to make 10 of these for the craft fair? If not, what can you do to make more products in the same time?*

- After 10 minutes, answer any questions students have. Remind students that they must do their group jobs in addition to any assembly job they have.

- Set the timer for 45 minutes of product assembly. Circulate around the room. Ask questions such as:

  *How are you making sure you'll have enough items made for the craft fair?*

  *Have you run into any problems with assembling the product, or with each other? How are you working to solve them?*

- Groups should keep working for the full 45 minutes or until they run out of materials to assemble their product, whichever comes first. After 45 minutes, all groups stop working, even if they have leftover materials or partially completed products. Have groups gather their products together and set aside any unused materials or incomplete items.

- Store all products and the Materials Order Forms until Session 2.

**Build**
*Session 2*
*40 minutes*

- Return the products and Materials Order Forms to each group. Tell students that all businesses have paperwork and their order form was just the beginning. Remind them that when they did the lemonade stand word problem, they first found the cost per serving of the lemonade, including the cost of both materials and labor. Then they decided the price to charge for a serving of lemonade.

- Explain that they need to figure out the cost of materials and labor for one item of the product they just made, before they can set a price for it. To find out the cost of materials, each group needs to complete lines E and F on the Materials Order Form. If any group points out that they have leftover materials that they did not make into products, then explain that these leftovers are still part of the materials cost. The only way they can get money back to cover the cost of the leftovers is by selling the finished products—so the total cost of materials has to be divided by the number of finished items.

- When all groups have finished calculating line F, distribute an Income Statement sheet to each Recorder. Have each group calculate the Labor cost per item by completing lines H and I. Then have them calculate the total cost per item by completing line J.

- Point out that each group needs to decide on a price for their product. Their price must cover the costs of materials and labor for each item, and provide a profit that meets the minimum listed in the Criteria. But they also need to set a price that customers are willing to pay for the product. Give groups up to 5 minutes to discuss and decide on a price. Instruct them to write the price on line K of the Income Statement sheet. Instruct them to leave lines L, M, and N blank until after the fair.

- Next, distribute a Product Information sheet to each group and give them 15 minutes to complete it. The questions on this sheet help the groups summarize what makes their product appealing to customers; check whether their product is priced appropriately; and decide how to promote the product at the craft fair. As needed, review the terms *marketing* and *promotion*, so students understand that they need to come up with a way to make customers want to buy their product. Point out that they will have time to create signs and posters the day of the fair; they do not have to do it during this session.

- When students have finished their paperwork, have groups store the products, Materials Order Forms, Income Statements, and Product Information sheets in a safe place until it is time to present at the craft fair.

### 5 Test & Present

*20 minutes prep, then 40 minutes for the craft fair*

- Ahead of time, confirm that the customers can join your class on the day of the craft fair.

- Be sure that you have prepared enough pretend money so that each customer may purchase at least one product, based on the prices that groups have set for their products.

- Assign each group a place in the room to set up their exhibit table, following the description in Constraints. Set out materials that are suitable for making signs and posters, such as construction paper, markers, and tape. Give groups 20 minutes to set up their display spaces.

- When everyone is ready, call in the customers. Suggest that they visit all the tables before making any purchases.

- Set the timer for 20 or 30 minutes, depending on the size of your class and the number of customers. The businesses have that much time to sell as much product as they can.

- When time is up, no more purchases may be made. Thank the customers for coming and escort them out of the craft fair with their purchases.

- Have each group take out their Order Forms and Income Statements. Point out that these two sheets include all the information about how much they've spent to produce their product. Now, they need to figure out how much money they took in from selling the product (income), and then figure out whether the business experienced a profit or a loss. You may wish to walk groups through completing their sheets as a class.

- When groups have finished their calculations, invite Speakers to report on the profit or loss of their new businesses. Was their business successful— that is, was it profitable? If not, what could they have done to change that? If it was, then how could they make the business even more profitable?

- When all groups have shared their results, give each student a Challenge Reflection sheet to complete.

## *Opportunities for Differentiation*

**To make it simpler:** Assemble different materials kits for student groups to choose from, each with directions for making a specific product and enough materials to make up to 20 items. Give each kit a flat rate, rather than requiring students to calculate separate costs for each material. Students base their per-item cost on the cost of the kit plus the flat rate for labor divided by the number of items the students make in 45 minutes using the kit.

**To make it harder:** Challenge students to include the material and labor costs associated with marketing and sales and still make a profit. Calculate labor based on a per-person, per-minute rate, rather than a flat rate. Allow each company to have two different products, and then compare how well each product sells.

# Lemonade Stand Word Problem

You are opening a lemonade stand. The lemonade recipe is below. Lemon juice costs $4.80 for 1 bottle (6 cups). Sugar is $0.36 per cup. Water is free. You already own the pitcher, spoon, measuring cup, and the table and chairs for the stand. Paper cups and straws cost $0.01 each. It costs $0.10 per second to pay all the workers to make the lemonade.

**What is the cost to make one serving of lemonade?**

Show your work and explain your thinking.

---

# Lemonade Recipe
## (makes about 12 servings, 6 ounces each)

1½ cups lemon juice

8 cups water

2 cups white sugar

1. In a pitcher, combine lemon juice and water.
2. Gradually stir in sugar until all sugar is added and completely dissolved.
3. Pour lemonade into paper cups. Add 1 straw to each cup.

# New Business Product Materials Price List

### 1¢ each
- Beads
- Buttons
- Pom-poms
- Small foam cutouts

### 3¢ each
- Bobby pins
- Clear plastic cups
- Hair clips
- Pipe cleaners
- Key rings
- Craft sticks
- Erasers
- Straws
- Cardboard tubes

### 3¢ per foot
- Yarn
- String

### 5¢ per foot
- Duct tape

### 5¢ per sheet
- Tissue paper
- Colored cellophane

### 7¢ per sheet
- Craft foam

Names    _____    _____

           _____    _____

# New Business Product Information

1.  What is the name of the product?   _____

2.  How is the product used?

     _____

3.  Why would someone want this product?

     _____

4.  What is the price of the product?   _____

5.  Does the price include a profit of at least 20 cents per item?   _____

6.  How does your business plan to promote the product at the craft fair?

     _____

     _____

Names _____    _____

_____    _____

# New Business Materials Order Form

Name of Product _____

| Material | A. Quantity (number of items or number of feet) | B. Cost (per item or per foot) | C. Total cost of this material (A × B for each material) |
|---|---|---|---|
|  |  |  |  |
|  |  |  |  |
|  |  |  |  |
|  |  |  |  |
|  |  |  |  |
|  |  |  |  |
|  |  |  |  |
|  |  |  |  |
|  |  |  |  |
| **D. Total cost of all materials** (sum of column C) |  |  |  |
| **E. Number of items assembled** (complete after Build) |  |  |  |
| **F. Cost per item assembled** (D ÷ E) (complete after Build) |  |  |  |

Names _____    _____

_____    _____

# New Business Income Statement

## Cost per Item and Selling Price

| | |
|---|---|
| **G. Total labor cost** | $5.00 |
| **H. Number of items produced** | |
| **I. Labor cost per item produced** (G ÷ H) | |
| **J. Total cost per item** (F + I)<br>(Use F from Materials Order Form.) | |
| **K. Selling price per item**<br>(Group decides. Must have a profit of at least $0.20.) | |

## Profit or Loss

| | |
|---|---|
| **L. Number of items sold at fair** | |
| **M. Total income** (K × L) | |
| **N. Total cost of items, sold and unsold** (D + G)<br>(Use D from Materials Order Form.) | |
| **Total profit or loss** (M − N) | |

How could you increase the business's profits? Describe two ways.

_____

_____

# Solar Water Heaters

## Curriculum Connections

Science: energy transfer, alternative energy sources
Social Studies: people & the environment, Earth Day

## Criteria for Product

- Must rely on sunlight to heat the water
- Must use materials that absorb heat energy from sunlight
- Must keep water warm when taken out of direct sunlight

## Constraints for Challenge

- May use only materials from the kit
- Must be able to hold 1 cup of water
- Must complete each stage in the time allowed

### Challenge

Summertime brings fun in the sun! Sunlight is also an important energy source. Students design, build, and test solar water heaters.

## Materials

### Solar Water Heater Materials (for each group)

- 1 empty plastic bottle, 1- or 2-L size, with top cut off
- 2 empty soda cans, 12-oz. size
- 1 cardboard box, shoebox size or larger
- 3 ft. of aluminum foil
- 3 × 1-ft. sheet of shiny, silver plastic gift wrap or emergency blanket (optional)
- 3 ft. of waxed paper
- 3 ft. of plastic wrap
- 2 paper plates
- 1 plastic shopping bag
- 5 paper towels
- 3 sheets of black construction paper
- 3 sheets of newspaper
- 20 cotton balls
- 1 yd. of masking tape
- 1 shopping bag (to hold materials kit)

### Tools

- Scissors
- Hole punches
- Funnels (tops cut from the bottles work well)
- Measuring cup
- Penknife (teacher use only)

### Additional Materials

- Extra kit materials for use during Investigate
- Chart paper
- Crayons
- Markers
- Pencils
- Plain paper
- Thermometers, science-kit type, 1 for each group
- Digital timers or stopwatches, 1 for each group
- Water in containers to carry outdoors (1 cup for each group)
- Computer with Internet access (optional)
- Projector for computer (optional)
- Examples of online infomercials, bookmarked (optional)
- Digital video cameras (optional)

# Before You Begin

- Choose an appropriate time and place for Test & Present. Solar water heaters work best in late spring and early summer, when the sun is highest in the sky. On a daily basis, the sun is highest in the sky between 11 a.m. and 3 p.m. during daylight saving time and between 10 a.m. and 2 p.m. during standard time. Locate a place outdoors that has full, direct sunlight for 30 minutes during this time frame, and that is out of the way of other classes. If you plan to have students perform infomercials, be sure there is also space for skit rehearsals nearby.

- Assemble a materials kit for each group, following the Solar Water Heater materials list.

- Cut the tops off the 2-liter bottles.

- The day before Test & Present, fill the containers you will use to carry water and let them sit at room temperature overnight. This ensures that the water used during Test & Present is the same temperature as the air, which produces clearer results.

- Make a set of Job Cards (page 13), a copy of the Test Results sheet (page 171), and a copy of Common Features of Infomercials (page 170) for each group.

- Make a copy of the Challenge Reflection sheet (page 14) for each student.

- If you plan to have students create infomercials during Test & Present, search online for a few 1- to 3-minute examples. Preview infomercials for suitability and for inclusion of the features listed on page 170. Bookmark any that you plan to show.

- Set up one digital timer where students can check on it, for timing the stages.

- Display the Criteria and Constraints where students can see them.

# 5-Step Process

### 1 Investigate
*Session 1*
*20 minutes*

- Provide each student with a sheet of plain paper and crayons. Instruct students to draw pictures that include the sun. Do not give them any other directions. The only requirement is that the picture contains the sun. Allow 10 minutes for drawing. If some students finish early, suggest they add more details.

- After 10 minutes, invite each student to find a partner and share their pictures. Each person should talk about the contents of her picture and what she was thinking when she drew it. Set the timer for 1 minute and tell the first person to talk for the entire minute without stopping. After 1 minute, reset the timer and have the second person take his turn talking for a full minute.

- Once sharing is complete, begin a class discussion about the similarities students noticed about their drawings. Many drawings will show summer fun, outdoor activities, bright colors, living things thriving, people dressed in lightweight clothing, and similar subjects.

- Ask students to consider why thoughts of the sun brought up these images. What is it about the sun that leads people to think like this? Guide students to talk about the fact that the sun gives heat and light.

- Invite students to think about ways living things use heat and light from the sun. Encourage them to consider both natural and manufactured uses. Ideas might include plants using light for photosynthesis, animals using sunlight to see or to stay warm, windows and sunroofs in houses letting in natural light, and sunlight heating Earth's land and water.

- Ask students to think about what happens when they turn on the hot water faucet. Where does the hot water come from? How does it get so hot? If students do not know about water heaters, explain that wherever there is a hot water faucet, there is a way to heat the water. Most hot water heaters have a tank with a heater that runs on natural gas, oil, or electricity. All of these energy sources have to be paid for, and except for a very few sources of electricity, all have negative effects on the environment.

- Ask students to think about the source of heat they were discussing a few minutes ago. The sun's energy is free and it is a natural part of the environment. Tell students that they will design and build solar water heaters—that is, water heaters that use heat energy from sunlight.

**① Investigate**

*Session 2*

*45 minutes*

- Choose a bright, sunny day for Session 2. It does not need to be warm, just sunny. A few fair-weather cumulus clouds are okay.

- Remind students that for this Challenge they will be designing, building, and testing solar-powered hot water heaters.

- Introduce the criteria for this Challenge. Ask students to think about possible materials for building a solar water heater. What properties do these materials need in order to meet the criteria?

- Display the solar water heater materials in one group kit and have students gather around. Ask students to describe some of the properties they are already familiar with that might be useful for meeting the criteria. For example, they know that a plastic bottle can hold water and that foil reflects light.

- If students do not bring it up, ask which materials would be good for absorbing heat from sunlight, which would be good for concentrating sunlight in one place (such as a container of water), and which would be good for holding onto heat when taken out of direct sunlight. Call on students to suggest ways they could find out which materials have these properties. Guide them to think about taking the materials outdoors for testing in sunlight.

- Work with the students to develop a few simple tests to conduct on the materials. The test ideas should come from the students. If students cannot think of workable tests, guide them. Remind students that in a fair test or experiment, all things are the same except for one variable. Tests might include the following:
  - Place one black and one white sheet of paper side by side in direct sunlight for 10 minutes, and then touch them to compare their temperatures.

- Pour the same amount of water into three identical cups. Wrap each cup in a different material. Let the cups sit in direct sunlight for 10 minutes, and then use a thermometer to compare the temperature of the water.

- Pour the same amount of water into a can and a bottle. Set them in direct sunlight for 10 minutes, and then measure the temperature of the water in each container. Set the containers in a shady spot, and then measure the temperature again after 10 minutes.

- Divide students into small, temporary groups and assign a test to each one. Go outdoors, set up, and conduct the tests. If students came up with only one or two tests, then set up multiple versions of the same test so that all students are actively involved.

- After the tests are complete, ask students to describe how direct sunlight affected each of the materials. Prompt discussion with questions such as these:

  *Which materials were warmer after sitting in direct sunlight? Which materials stayed about the same temperature?*

  *Do you think the color of a material affects the way it absorbs and holds heat from sunlight? What about its shape or size?*

  *Do some materials cool down faster than others when brought out of direct sunlight?*

  *Which heats up more quickly in direct sunlight, a closed container or an open container?*

  *Do you think the results might change if you left the materials in sunlight longer?*

  *Do you have any other questions about this Challenge?*

- Review the Criteria and Constraints for this Challenge, as well as the time limits for Brainstorm, Plan, and Build.

**2** ▶ *Brainstorm*
*20 minutes*

- Set the stage by reminding students that Brainstorm is for writing down ideas without stopping to think about how well they will work.

- Give each student a sheet of plain paper. Instruct them to work alone to brainstorm ideas for a solar water heater made of the materials they investigated.

- Allow 10 minutes for students to work; as they do, walk around the room and look over their shoulders. Ask questions that guide students to think of new possibilities. Suggested questions:

  *Which materials can hold water? How much water does the heater need to hold?*

  *Which materials are good at absorbing heat in direct sunlight?*

  *Which materials could help direct sunlight toward the water?*

  *Which material are you using here? Why did you choose that material?*

  *How can you maximize the temperature increase, using that material?*

*What job does this material do? Are any other materials good for that job?*

*Will the water evaporate? How can you keep that from happening?*

*What else do you know about what happens to materials on a hot, sunny day? How do you know these things?*

**③ Plan**
*30 minutes*

- Instruct students to form groups of four (or three as needed). Stand back and let students form groups independently, in whatever way they choose. Students should sit down together once they've formed their groups.

- After groups have formed, hand a set of Job Cards to each group. They may choose jobs in any way they like as long as each person has at least one job, and all four jobs are assigned.

- Tell students that they will share and discuss all of their ideas from Brainstorm, not just their favorites. The group should share and discuss all ideas before deciding on anything for the group plan. Set the timer for 20 minutes.

- Visit with each group as they work. Compliment good collaborative behaviors. Challenge students to stretch their thinking, use what they know about how materials respond to sunlight, and be willing to combine ideas from more than one person. Suggested questions:

    *How did you work together to create this plan?*

    *Which part of this design holds the water?*

    *What material will absorb heat from sunlight?*

    *How did you choose ideas to use for the group plan?*

    *Which part of this design keeps the water from cooling down?*

    *Why did you choose this material? Did the way it performed during Investigate affect your decision to use it?*

- As each group settles on a design, give each person a new sheet of plain paper. Tell them that each group member should draw and label a diagram of the solar water heater. All diagrams within the group need to match. This shows that each person understands the plans for the solar water heater in the same way. If there are differences, then the group should discuss them and figure out what the working plan is.

**④ Build**
*65 minutes*

- Tell students it is time to build the solar-powered water heaters.

- As a class, review the procedure for building. Call on student volunteers to list the steps in order, one at a time.

    - Help to build and do your group job.

    - Follow the plan as you build.

    - Change the plan only if there is a good engineering reason to do so—for example, the plan isn't working as expected.

    - Notify the teacher and explain the reason for any change.

    - Make changes to the original plan in a new color, or draw up a new one and label it Plan 2.

- Invite the Materials Managers to pick up the materials kits. Set the timer for 60 minutes.
- Move around the room, checking on students' work. Ask questions that require students to explain their thinking and rely on their observations. Sample questions:

  *Is it coming together the way you expected? If not, what will you do?*

  *How can you make sure the water gets as much sunlight as possible?*

  *How will you know if your heater works when you test it?*

  *How did you divide the work for this project?*

- Store the completed solar water heaters in a safe place until it is time for Test & Present.

## 5 ▸ Test & Present

*15 minutes to introduce the infomercial (optional); 10 minutes to set up; 30 minutes to test; 20 minutes total to present*

- If desired, introduce the concept of infomercials. Display or read aloud the features that most infomercials have in common, such as those listed on page 170. Show the examples you located and bookmarked. Ask students to identify the features of infomercials as they watch the examples.
- Tell students it's time to test their solar water heaters. Go outdoors with the water heaters, thermometers, digital timers, Test Results sheets, and room-temperature water. If you plan to create infomercials, also bring digital cameras, paper, and markers.
- Set up the solar water heaters in the place chosen during Before You Begin. Measure and pour 1 cup of water into each water tank. Instruct the Timekeeper to measure the starting temperature of the water and the Recorder to write the starting temperature on the Test Results chart under "0 minutes."
- Tell the Timekeepers to set the timers for 5 minutes. The Timekeepers and Recorders are in charge of checking and recording the temperature of the water in the heaters every 5 minutes, and re-setting the timers, for a total of 30 minutes. During this time, each group will also be working on an infomercial for their solar water heater. Infomercial practice continues even during temperature checks.
- Loan each group a digital video camera so they can record their infomercials. Caution groups to be careful not to spill their heaters while recording infomercials, because they still need to measure and record the temperature of the water, even after they go indoors.
- When all students have finished recording—temperatures and infomercials—go back indoors with the solar water heaters.
- Tell Timekeepers and Recorders they have one more measurement to make. Now that the water is out of the sunlight, it is time to find out how well the solar water heater tanks hold the heat. Set the timer for 5 minutes so the Timekeepers and Recorders know when to take this measurement.

- After the indoor measurement has been recorded, allow each group to have a turn presenting. First show the infomercial, and then have students share the results on their Test Results sheet. Last, have the class ask three questions of each group:

    *Does the solar water heater meet the criteria?*

    *Are there any differences between what the solar water heater does and what the infomercial says it does?*

    *Does the infomercial tell about any problems with the solar water heater?*

- Provide a copy of the Challenge Reflection sheet for each student.

## *Opportunities for Differentiation*

**To make it simpler:** Before beginning, spray-paint the outsides of the bottles and cans black and omit the black paper from the materials kit. Provide some basic directions for building a solar water heater. Tell students to line the inside of the box with a material that reflects sunlight, place the black-painted container in the box, and then add insulating materials around the container.

**To make it harder:** Omit either the box or the tape from the materials kit. Instead, include rubber bands, string, and pipe cleaners.

# Common Features of Infomercials

An *infomercial* is a television commercial for an unusual product. Some infomercials are just a few minutes long. Others are as long as a whole television show. Many short infomercials have the following features in common.

1. Words are written in very large letters.
2. The narrator is overly dramatic.
3. They show a problem first, or a few related problems.
4. They ask the viewers if they have had these problems.
5. THEN, they show the product. They explain how it fixes the problems.
6. They use persuasive language.
7. They offer extra bonus items if the viewer acts now.
8. They insist that the viewer needs to call right away to get the super offer.

Names _____    _____

_____    _____

# Solar Water Heater Test Results

**Water Temperature**

| Time (min.) | 0 | 5 | 10 | 15 | 20 | 25 | 30 |
|---|---|---|---|---|---|---|---|
| Temp (°F) | | | | | | | |

Water temperature 5 minutes after heater was taken out of sunlight _____°F

1. Which material(s) absorb heat from sunlight, and why did you choose them?

   _____

   _____

2. Which materials keep the water from losing heat, and why did you choose them?

   _____

   _____

3. Is the water heater successful at heating water? Use your test results to support your answer.

   _____

   _____

4. Is the water heater successful at keeping the water warm? Use your test results to support your answer.

   _____

   _____

# Roller Coasters

## Curriculum Connection

Science: potential (stored) & kinetic energy, variables

## Criteria for Product

Roller coasters will be scored on:

- Number of turns in track
- Number of hills in track
- Number of upside-down loops in track
- Total distance covered, measured along the base
- Number of completed runs by a marble

## Constraints for Challenge

- Only materials found in the classroom or brought in for this Challenge may be used
- Must complete each stage in the time allowed
- Marble and coaster may not be touched or bumped after marble is released for its run
- Limit of three attempts to run a marble on the track

### Challenge

Summertime means cotton candy and carnival games. For those who dare, it means thrilling amusement-park rides! Students design, build, and test roller coasters.

## Materials

### Roller Coaster Materials (for the class)

- Assorted plastic bottles and jars
- Assorted sticks from trees
- Paper towel and bath tissue tubes
- Empty potato-chip cans
- 20 dowels or pieces of PVC pipe, any lengths
- 100 ft. of vinyl or rubber tubing or foam pipe insulation, wide enough for marbles to run through

- 100 paper or foam plates
- 100 foam cups
- Paper
- A few rolls of clear tape
- A few rolls of masking tape

### Tools

- Rulers
- Tape measures
- Scissors
- Hole punches

### Additional Materials

- Chart paper
- Markers

- Pencils
- Plain paper
- Digital timer
- Marbles
- Several marble run toys (optional)
- Computer with Internet access (optional)
- Computer projector (optional)
- Computers with Internet access for student use (optional)
- Document camera (optional)

**Note:** *Provide roller-coaster materials in quantities that are readily available.*

# Before You Begin

- Request donations of recyclables from families.
- Gather printed versions of a variety of amusement-park maps, including parks that students may be familiar with. Either contact the amusement parks ahead of time to request brochures, or locate and print maps from park websites. A dozen or more will provide a good variety. Bookmark some online maps for class discussion, if desired.
- Make a set of Job Cards (page 13) and a copy of Test Results (page 178) for each group.
- Make a copy of the Challenge Reflection sheet (page 14) for each student.
- Locate and bookmark a roller-coaster simulator on the Internet, if you plan to use one. A search for "roller-coaster simulator" (or "simulation") returns good results.
- Set up the digital timer where students can check on it.
- Display the Criteria and Constraints where students can see them.

# 5-Step Process

**1** **Investigate**
*Session 1*
**40 minutes**

- Pass out the amusement-park maps so that each student or pair of students has one. Tell students to look for the roller coasters on the maps and to compare theirs with the coasters on other students' maps. Allow up to 10 minutes for this.
- Display one park map for the class to see, by either placing it under a document camera or displaying the website that shows the map. Ask students to share which roller coasters are their favorites and why. Repeat the process with several maps.
- Invite students to think about roller coasters from a science perspective. Ask students:

  *How do you think roller coasters work?*

  *What makes the car stay on the track when it turns? What about when the track makes the car turn upside down in a loop?*

  *Where is the highest part of a roller coaster? Is this true for all roller coasters?*

  *What do the ends of roller-coaster tracks have in common?*

- Explain that although today's roller coasters are motorized and include many safety devices that keep the cars on the track and the people in the cars, they share some basic science principles with the earliest roller coasters that rolled down mountainsides: Roller coasters convert potential energy (stored energy) into kinetic energy (energy of motion). You cannot get more kinetic energy out of a roller-coaster car during its trip than you put into it at the beginning.

- Display the roller-coaster simulator website you bookmarked earlier for class discussion, or arrange for students to use this site through computers with Internet access. Explain that when engineers design anything, including roller coasters, they often use computer simulators to test ideas before building. Students will use a simulator to practice designing roller coasters before they build one for real.

- Allow about 20 to 30 minutes for practicing roller-coaster design. As they design, remind students to notice how different features affect the performance of a roller coaster.

**1 Investigate**
**Session 2**
**45 minutes**

- As a class, discuss what students learned about roller coasters from the simulator. Did every roller coaster design work? How do you know? What features do successful roller-coaster designs have in common? What features do failing designs have in common? Encourage students to describe the success or failure in terms of energy: Which features give the roller-coaster car enough potential energy to make it through the whole track? Which features cause the car to stop rolling partway through, or to crash?

- Ask students to think about a roller-coaster car that stops before it finishes the track. Invite students to share ideas for solving this problem. Guide students to conclude that the car needs to have enough potential energy from the first hill to go up later hills, as well as through loops and turns.

- Tell students that it's time to test their ideas from the simulator in the real world. Bring out the marble run toys, if available. Explain that the marble runs will provide more practice for their engineering Challenge: to design and build roller-coaster runs for marbles, using the materials they think will work best.

- Give students time to explore and play. Let students know that even though they may be having fun, they need to pay attention to testing their ideas from the simulator. When they design their own roller coasters, they will be combining everything they learned on the simulator, plus what they are learning now from the marble run.

- Call on students to share what they observe as they explore. Ask questions to encourage observations. For example:

  *How can you make sure a marble has enough energy to go uphill and around loops and turns?*

  *Is it possible for a marble to go over a hill that's taller than the one it starts on?*

  *Does distance make a difference in how well the marble run works?*

  *How can you slow the marble at the end of the run?*

- Encourage students to discuss how they could use their observations of the marble run toy in the design of their marble roller coasters. If your students are familiar with the terms *potential energy* and *kinetic energy*, then reinforce the correct use of these terms during the discussion. However, knowing the terms is not critical to this Challenge. It is more important for students to observe patterns in how the design of the marble track affects the motion of the marble, and to be able to apply those observations to the design of their roller coasters.

- Tell students that next, they'll be designing their roller coasters. Review with students the Criteria and Constraints, as well as the time limits for Brainstorm, Plan, and Build. Show them the Test Results sheet so they know how their finished roller coasters will be scored.

## **2** Brainstorm
### 15 minutes

- Place the roller-coaster materials randomly around the classroom where students can see them, but do not display them all together in one place. Do this when no students are in the room.

- Tell students that for this Challenge, they'll brainstorm as a class. Everyone will be able to see the same ideas list from Brainstorm during Plan.

- Explain to students that it's up to them to choose materials for their roller coasters. Encourage students to look around the classroom for materials they could use. (Be prepared to veto any item you do not wish them to use.) If they think of something they want to bring in from home, that's okay too, as long as they have the permission of the adults in charge.

- Ask students to think about the parts of a roller coaster. Record the ideas prompted by the questions. Suggested questions:

  *How can you make the roller coaster turn?*

  *What materials could you use for a track?*

  *How can you build each part, using the available materials?*

  *Should the marble run through tubes, on a track, or a little of both?*

  *Can you include an upside-down loop in your roller coaster? Where should you place it? What materials could you use to build a loop?*

  *How can you change the shape of that material to make it work for a roller coaster?*

- As students list ideas, write them on chart paper or use another method that all students can see, and that they can refer to during the rest of the Challenge.

## **3** Plan
### 30 minutes

- Have students form groups of four in whatever way they choose. For this Challenge, there are no requirements for grouping other than having four (or three) people to a group.

- Once groups are formed, give each one a set of Job Cards. Students may use any method of choosing jobs as long as each person has at least one job and all four jobs are assigned. Step in and facilitate only if necessary.

- Provide each group with chart paper, markers, and pencils for sketching out their plan. Set the timer for 20 minutes.

- Tell students that for this Challenge, all the students in a group work together on one version of the plan. They should take turns drawing. Their plan should include rough measurements and a description of how the roller coaster will work, and should use color coding to show each different material.

- Walk around the room, checking on group progress. Remind students of the Criteria and ask how their designs address each item on the list. Suggested questions:

  *How does your roller coaster work?*

  *Where is the tallest hill? Why are you placing it there?*

  *How long will your roller coaster be?*

  *Will your roller coaster be all in one straight line or will it have turns?*

  *What makes the marble stop at the end?*

  *Where will these materials come from?*

  *How will you hold these two materials together?*

- Make sure all groups have agreed to a plan before proceeding to Build.

**Build**

*60 minutes*

- Let students know that it's time to build the roller coasters. Allow 2 minutes for groups to review the procedure for Build. Drop in on conversations to be sure students are mentioning all the important steps:

  - Follow the plan as it's written.
  - If the plan changes as you build, tell the teacher why and record the changes in a new color on the plan, or draw up a Plan 2.
  - The Build stage ends at the same time for all groups, even groups that made a change.
  - Work together and be sure everyone does their job and helps to build.

- Once students have reviewed the Build procedure, set the timer for 50 minutes. Call on the Materials Managers to collect any needed materials. Provide each group with a marble for testing their coasters as they build.

- Circulate around the room, chatting with each group. Lead students to their own conclusions by asking guiding questions, such as:

  *What is everyone's job in this group? What is everyone's role in building the roller coaster?*

  *How are you holding the parts together? Can the marble get over that seam?*

  *Are you testing as you go? Does the marble make it all the way through without any help?*

  *Have you had any surprises, or anything that didn't work as planned? What have you done about it?*

- Store the roller coasters in a safe place until it is time for Test & Present.

## ⑤ Test & Present

*10–15 minutes per group*

- Clear an area of floor or large tabletop as a test area. Call groups to present, one at a time.
- Provide a Test Results sheet on which the Recorder can write all scores and results.
- Invite the Speaker to explain each part of the group's roller coaster and demonstrate the path the marble takes, without releasing the marble. The Speaker should include reasons for the design decisions, and address how the group considered potential and kinetic energy in their design. Allow the class time to ask questions.
- Tell the group they have three tries to get the marble to complete the run and stop at the end. As needed, suggest that the Materials Manager, Timekeeper, and Recorder each have one turn. Be sure that no one is in a position to shake the roller coaster or the table it may be resting on.
- Follow the same procedure for the other groups. When all groups have tested and scored their roller coasters, discuss and compare results as a class.
- Distribute Challenge Reflection sheets for students to complete.

## Opportunities for Differentiation

**To make it simpler:** Provide a specific list of materials. Show students some ways to use the materials to build a roller coaster. Demonstrate a design or two.

**To make it harder:** Add the speed of the marble to the criteria. Have students use stopwatches to time the marble's run in seconds, and then divide the distance of the track by the time to find the marble's speed in inches per second.

Names _____    _____

_____    _____

# Roller Coaster Test Results

| Criterion | Score |
|---|---|

### 1. Turns
Score 1 point for each turn

_____

### 2. Hills
Score 2 points for each hill, not counting the first hill

_____

### 3. Loops
Score 3 points for each upside-down loop

_____

### 4. Distance
Measure distance from start to end at base of roller coaster.
Score 1 point for each foot, rounded to the nearest foot.

_____

### 5. Completed runs
Score 5 points for each completed run, up to three runs.

_____

### Total Score

_____

# Appendices

# References

Barell, John. 2006. *Problem-Based Learning: An Inquiry Approach*. Thousand Oaks, CA: Corwin Press.

Boaler, Jo. 1998. "Open and Closed Mathematics: Student Experiences and Understandings." *Journal for Research in American Mathematics Education,* 29 (1), 41.

Bransford, John D., Ann L. Brown, and Rodney R. Cocking, ed. 2000. *How People Learn: Brain, Mind, Experience, and School: Expanded Edition*. Washington, D.C.: National Academies Press.

Drew, David E. 2011. *STEM the Tide: Reforming Science, Technology, Engineering, and Math Education in America*. Baltimore, MD: The Johns Hopkins University Press.

Helm, Judy Harris, and Lillian Katz. 2001. *Young Investigators: The Project Approach in the Early Years*. New York: Teachers College Press.

Jacobs, Heidi Hayes. 2010. *Curriculum 21: Essential Education for a Changing World*. Alexandria, VA: Association for Supervision and Curriculum Development (now ASCD).

Lantz, Hayes Blaine, Jr. 2009. "STEM Education: What Form? What Function?" *SEEN* Magazine 11 (2): 28–29.

Larmer, John, and John R. Mergendoller. 2010. "7 Essentials for Project-Based Learning." *Educational Leadership* 68 (1).

MacDonell, Colleen. 2006. *Project-Based Inquiry Units for Young Children: First Steps to Research for Grades Pre-K–2*. Santa Barbara, CA: Linworth Libraries Limited.

Nagel, Nancy G. 1996. *Learning Through Real-World Problem Solving: The Power of Integrative Teaching*. Thousand Oaks, CA: Corwin Press.

Schunn, Christian D. 2009. "How Kids Learn Engineering: The Cognitive Science Perspective." *The Bridge: Linking Engineering and Society* 39 (3): 32–37.

Wiggins, Grant, and Jay McTighe. 2005. *Understanding by Design, 2nd ed.* Alexandria, VA: Association for Supervision and Curriculum Development (now ASCD).

## Organizations & Initiatives

American Society for Engineering Education (ASEE)
eGFI (Engineering—Go For It!)
1818 N Street NW, Suite 600
Washington, DC 20036
www.egfi-k12.org

ASCD (formerly Association for Supervision and Curriculum Development)
1703 N. Beauregard Street
Alexandria, VA 22311-1714
www.ascd.org

The Coalition for Science After School
University of California
Lawrence Hall of Science, #5200
Berkeley, CA 94720-5200
www.afterschoolscience.org

Common Core State Standards Initiative
www.corestandards.org

International Society for Technology in Education
1710 Rhode Island Avenue NW, Suite 900
Washington, DC 20036
www.iste.org

International Technology and Engineering Educators Association
1914 Association Drive, Suite 201
Reston, VA 20191-1539
www.iteea.org

National Science Teachers Association
1840 Wilson Boulevard
Arlington, VA 22201
www.nsta.org

Next Generation Science Standards
Achieve, Inc.
1400 16th Street NW, Suite 510
Washington, DC 20036
www.nextgenscience.org

The Partnership for 21st Century Skills
1 Massachusetts Avenue NW, Suite 700
Washington, DC 20001
www.p21.org

Teach Engineering: Resources for K–12
www.teachengineering.org

TryEngineering
www.tryengineering.org

# Content & Skills Alignment Charts

| Next Generation Science Standards: A Framework for K–12 Science Education | Engineering Challenges | | | | | | | | | | | | | | | | | | | |
|---|---|---|---|---|---|---|---|---|---|---|---|---|---|---|---|---|---|---|---|---|
| | 1 Rockets | 2 Catapults | 3 Compasses | 4 Healthy Snacks | 5 Simple Shelters | 6 Skyscrapers | 7 Model Cars | 8 Board Games | 9 Ski Lifts | 10 Winter Coats | 11 Greenhouses | 12 Bird Feeders | 13 Parachutes | 14 Bridges | 15 Land-Reuse Models | 16 Mining Tools | 17 Oil-Spill Cleanups | 18 New Businesses | 19 Solar Water Heaters | 20 Roller Coasters |
| PS2.A: An object at rest typically has multiple forces acting on it, but they add to give zero net force on the object. Forces that do not sum to zero can cause changes in the object's speed or direction of motion. | X | X | | | | X | X | | X | | | | X | X | | | | | | X |
| PS2.B: Objects in contact exert forces on each other (friction, elastic pushes and pulls). Electric, magnetic, and gravitational forces between a pair of objects do not require that the objects be in contact—for example, magnets push or pull at a distance. | X | X | X | | | X | X | | X | | | | X | X | | | | | | X |
| PS3.A: The faster a given object is moving, the more energy it possesses. | X | X | | | | | X | | X | | | | | | | | | | | X |
| PS3.B: Light . . . transfers energy from place to place. For example, energy radiated from the sun is transferred to Earth by light. When this light is absorbed, it warms Earth's land, air, and water and facilitates plant growth. | | | | | | | | | | | X | | | | | | | | X | |
| PS3.B: Energy is transferred out of hotter regions or objects and into colder ones by the processes of conduction, convection, and radiation. | | | | | | | | | | X | X | | | | | | | | | |
| PS3.D: The expression "produce energy" typically refers to the conversion of stored energy into a desired form for practical use. . . . Food and fuel . . . release energy when they are digested or burned. | X | X | | X | | | X | | X | X | X | | | | | | | | X | X |

| Next Generation Science Standards: A Framework for K–12 Science Education | 1 Rockets | 2 Catapults | 3 Compasses | 4 Healthy Snacks | 5 Simple Shelters | 6 Skyscrapers | 7 Model Cars | 8 Board Games | 9 Ski Lifts | 10 Winter Coats | 11 Greenhouses | 12 Bird Feeders | 13 Parachutes | 14 Bridges | 15 Land-Reuse Models | 16 Mining Tools | 17 Oil-Spill Cleanups | 18 New Businesses | 19 Solar Water Heaters | 20 Roller Coasters |
|---|---|---|---|---|---|---|---|---|---|---|---|---|---|---|---|---|---|---|---|---|
| **Engineering Challenges** | | | | | | | | | | | | | | | | | | | | |
| LS1.C: Animals and plants alike generally need to take in air and water; animals must take in food and plants need light and minerals. | | | | X | | | | | | | X | X | | | | | | | | |
| LS3.B: The environment also affects the traits that an organism develops—differences in where they grow or in the food they consume may cause organisms that are related to end up looking or behaving differently. | | | | | | | | | | | X | X | | | | | | | | |
| ESS2.A: Human activities affect earth's systems and their interactions at its surface. | | | | | X | X | | | | | | | | | X | X | X | | | |
| ESS2.C: Water continually cycles among land, ocean, and atmosphere via transpiration, evaporation, condensation and crystallization, and precipitation, as well as downhill flows on land. | | | | | | | | | | | X | | | | | | | | X | |
| ESS3.A: All materials, energy, and fuels that humans use are derived from natural sources and their use affects the environment in multiple ways. | | | | | X | | | | | | | | | | X | X | | | X | |
| ESS3.C: Human activities in agriculture, industry, and everyday life have had major effects on the land, vegetation, streams, ocean, air, and even outer space. But individuals and communities are doing things to help protect Earth's resources and environments. | | | | | X | X | | | | | | | | | X | X | X | X | X | |
| ETS1.A: Possible solutions to a problem are limited by available materials and resources (constraints). The success of a designed solution is determined by considering the desired features of a solution (criteria). Different proposals for solutions can be compared on the basis of how well each one meets the specified criteria for success or how well each takes the constraints into account. | X | X | X | X | X | X | X | X | X | X | X | X | X | X | X | X | X | X | X | X |

| Next Generation Science Standards: A Framework for K–12 Science Education | Engineering Challenges | | | | | | | | | | | | | | | | | | | |
|---|---|---|---|---|---|---|---|---|---|---|---|---|---|---|---|---|---|---|---|---|
| | 1 Rockets | 2 Catapults | 3 Compasses | 4 Healthy Snacks | 5 Simple Shelters | 6 Skyscrapers | 7 Model Cars | 8 Board Games | 9 Ski Lifts | 10 Winter Coats | 11 Greenhouses | 12 Bird Feeders | 13 Parachutes | 14 Bridges | 15 Land-Reuse Models | 16 Mining Tools | 17 Oil-Spill Cleanups | 18 New Businesses | 19 Solar Water Heaters | 20 Roller Coasters |
| ETS1.B: Research on a problem should be carried out—for example, through Internet searches, market research, or field observations—before beginning to design a solution. | X | X | X | X | X | X | X | X | X | X | X | X | X | X | X | X | X | X | X | X |
| ETS1.B: An often productive way to generate ideas is for people to work together to brainstorm, test, and refine possible solutions. Testing a solution involves investigating how well it performs under a range of likely conditions. Tests are often designed to identify failure points or difficulties, which suggest the elements of the design that need to be improved. | X | X | X | X | X | X | X | X | X | X | X | X | X | X | X | X | X | X | X | X |
| ETS1.B: At whatever stage, communicating with peers about proposed solutions is an important part of the design process, and shared ideas can lead to improved designs. | X | X | X | X | X | X | X | X | X | X | X | X | X | X | X | X | X | X | X | X |
| ETS1.B: There are many types of models, ranging from simple physical models to computer models. They can be used to investigate how a design might work, communicate the design to others, and compare different designs. | X | X | X | | X | X | X | | X | X | | X | X | X | X | X | | | | X |
| ETS1.C: Different solutions need to be tested in order to determine which of them best solves the problem, given the criteria and the constraints. | X | X | X | X | X | X | X | X | X | X | X | X | X | X | X | X | X | X | X | X |
| ETS2.A: Knowledge of relevant scientific concepts and research findings is important in engineering. | X | X | X | X | | | X | | X | X | X | X | X | | | | X | | X | X |
| ETS2.B: Engineers improve existing technologies or develop new ones to increase their benefits . . ., to decrease known risks . . ., and to meet societal demands. | X | X | X | X | X | X | X | X | X | X | X | X | X | X | X | X | X | X | X | X |

| 21st Century Learning Core Subjects and Student Outcomes | Engineering Challenges | | | | | | | | | | | | | | | | | | | |
|---|---|---|---|---|---|---|---|---|---|---|---|---|---|---|---|---|---|---|---|---|
| | 1 Rockets | 2 Catapults | 3 Compasses | 4 Healthy Snacks | 5 Simple Shelters | 6 Skyscrapers | 7 Model Cars | 8 Board Games | 9 Ski Lifts | 10 Winter Coats | 11 Greenhouses | 12 Bird Feeders | 13 Parachutes | 14 Bridges | 15 Land-Reuse Models | 16 Mining Tools | 17 Oil-Spill Cleanups | 18 New Businesses | 19 Solar Water Heaters | 20 Roller Coasters |
| ELA | X | X | X | X | X | X | X | X | X | X | X | X | X | X | X | X | X | X | X | X |
| Science | X | X | X | | | X | X | | X | X | X | X | X | X | X | X | X | | X | X |
| Math | | | | X | | X | X | X | X | X | | | X | | X | | | X | | |
| Arts | | | | | | | | X | | X | | | | | X | | | X | | |
| Creativity and innovation | X | X | X | X | X | X | X | X | X | X | X | X | X | X | X | X | X | X | X | X |
| Critical thinking and problem solving | X | X | X | X | X | X | X | X | X | X | X | X | X | X | X | X | X | X | X | X |
| Communication and collaboration | X | X | X | X | X | X | X | X | X | X | X | X | X | X | X | X | X | X | X | X |
| Information literacy | X | X | X | X | X | X | X | X | X | X | X | X | X | X | X | X | X | X | X | X |
| Media literacy | | | | | | | | | | | | | | | | | | X | X | |
| Information and communications technology literacy | X | X | X | X | | X | X | X | X | | | | X | | | | | X | X | X |
| Flexiblity and adaptability | X | X | X | X | X | X | X | X | X | X | X | X | X | X | X | X | X | X | X | X |
| Initiative and self-direction | X | X | X | X | X | X | X | X | X | X | X | X | X | X | X | X | X | X | X | X |
| Social and cross-cultural skills | X | X | X | X | X | X | X | X | X | X | X | X | X | X | X | X | X | X | X | X |
| Productivity and accountability | X | X | X | X | X | X | X | X | X | X | X | X | X | X | X | X | X | X | X | X |
| Leadership and responsibility | X | X | X | X | X | X | X | X | X | X | X | X | X | X | X | X | X | X | X | X |

| Common Core Standards for Mathematical Practice | Engineering Challenges | | | | | | | | | | | | | | | | | | | |
|---|---|---|---|---|---|---|---|---|---|---|---|---|---|---|---|---|---|---|---|---|
| | 1 Rockets | 2 Catapults | 3 Compasses | 4 Healthy Snacks | 5 Simple Shelters | 6 Skyscrapers | 7 Model Cars | 8 Board Games | 9 Ski Lifts | 10 Winter Coats | 11 Greenhouses | 12 Bird Feeders | 13 Parachutes | 14 Bridges | 15 Land-Reuse Models | 16 Mining Tools | 17 Oil-Spill Cleanups | 18 New Businesses | 19 Solar Water Heaters | 20 Roller Coasters |
| 1. Make sense of problems and persevere in solving them. | X | X | X | X | X | X | X | X | X | X | X | X | X | X | X | X | X | X | X | X |
| 2. Reason abstractly and quantitatively. | X | X | X | X | X | X | X | X | X | X | X | X | X | X | X | X | X | X | X | X |
| 3. Construct viable arguments and critique the reasoning of others. | X | X | X | X | X | X | X | X | X | X | X | X | X | X | X | X | X | X | X | X |
| 4. Model with mathematics. | X | X | | X | | X | X | X | X | | | | X | X | X | X | | X | X | X |
| 5. Use appropriate tools strategically. | X | X | X | X | X | X | X | X | X | X | X | X | X | X | X | X | X | X | X | X |
| 6. Attend to precision. | X | X | X | X | X | X | X | X | X | X | X | X | X | X | X | X | X | X | X | X |
| 7. Look for and make use of structure. | | | | X | | X | X | | | | | | X | X | X | X | X | X | X | X |
| 8. Look for and express regularity in repeated reasoning. | X | | | X | | X | | | X | | | X | X | | | | X | | X | X |